CABLE ADVERTISING

New Ways to New Business

Kensinger Jones
Michigan State University

Thomas F. Baldwin
Michigan State University

Martin P. Block
Northwestern University

Prentice-Hall, Englewood Cliffs, New Jersey 07632

Library of Congress Cataloging-in-Publication Data

Jones, Kensinger, (date)
Cable advertising.

1. Cable television advertising—United States.
I. Baldwin, Thomas F. II. Block, Martin P., (date)
III. Title.
HF6146.T42J65 1986 659.14'3 85-28305
ISBN 0-13-109828-4

This book is dedicated to the advertisers, advertising agencies, and system operators who are now pioneering new ways to new business through cable advertising.

Editorial/production supervision: Mary Bardoni and Erica Orloff
Interior design: Mary Bardoni
Cover design: Lundgren Graphics, Ltd.
Manufacturing buyers: Ed O'Dougherty and Carol Bystrom

Printed in the United States of America

10 9 8 7 6 5 4 3 2 1

ISBN 0-13-109828-4 01

Prentice-Hall International (UK) Limited, *London*
Prentice-Hall of Australia Pty. Limited, *Sydney*
Prentice-Hall Canada Inc., *Toronto*
Prentice-Hall Hispanoamericana, S.A., *Mexico*
Prentice-Hall of India Private Limited, *New Delhi*
Prentice-Hall of Japan, Inc., *Tokyo*
Prentice-Hall of Southeast Asia Pte. Ltd., *Singapore*
Editora Prentice-Hall do Brasil, Ltda., *Rio de Janeiro*
Whitehall Books Limited, *Wellington, New Zealand*

CONTENTS

PREFACE

A powerful new advertising force for national, regional, and local marketers is just beginning to come into its own. This new medium, cable television, evolved from something called "community antenna television" back in the 1950s. Originally intended to improve reception in fringe areas, CATV came to mean new sources of programming for viewers throughout the country. Today, in many places, subscribers have 35 channels or more to choose from and an amazing variety of programs directed to dozens of special interests.

For the advertiser, this means a new type of segmented audience, a greater chance than ever to expose *qualified* prospects to products and services through sight, sound, color, and action.

Advertising agencies and advertisers have the opportunity to develop media plans that zero in on those most likely to be interested in the item or service being sold. Creative people are finding new forms to use, freedom from the traditional constraint of just 30 seconds for a TV message.

Cable system operators are rapidly discovering what advertising can do for company revenues. Beyond that, some are finding that by working with local advertisers, they can initiate new and appreciated local programming.

Cable advertising is a unique business opportunity. Those who discover it and use it properly will have a strong competitive advantage. Those who ignore it do so at some risk.

A close look at what has happened, is happening, and is likely to occur in the foreseeable future should prove useful to those using or considering cable advertising.

A great many people helped in the making of this book. We thank them all, particularly the following:

Ann Alchin, our typist and organizer

Rita Aldridge, of the William R. Biggs/Gilmore Agency of Kalamazoo

Anthony Antin, *Reader's Digest*

Harold Bennett, Media Director for the William R. Biggs/Gilmore Agency of Kalamazoo

Kathy Boyal, Coaxial Communications, Columbus, Ohio

Richard Burkholder of Burkholder Flint Associates, Columbus, Ohio

Dave Gettys of Coaxial Communications, Columbus, Ohio

Bradley Greenberg, Chairman, Department of Telecommunication, Michigan State University

Tom Hammill, President, Universal Images Productions

Amy Hester of Burkholder Flint Associates, Columbus, Ohio

Saralee Hymen of the Cable Advertising Bureau for the materials from the bureau and her own suggestions

Don Keaton, Vice-President of Archway Cookies

Gary Krokker, Cable TV Sales Representative for Continental Cable of Lansing, Michigan

Delbert M. Lothes, BSA Industries, Columbus, Ohio

Steve McVoy of Coaxial Communications, Columbus, Ohio

Jane Meagher, Centel Videopath, Inc., Chicago, Illinois

J. W. Messner of the Messner Agency, Grand Rapids, Michigan

Ron Paugh, Ph.D. student, Michigan State University

Larry Redd, General Manager, LAN Cable Radio, Lansing, Michigan

Burl H. Robins, Creative Director of the Stevens Agency in Grand Rapids, Michigan

Sandy Rushton, Advertising Manager, Continental Cablevision, Lansing, Michigan

Chris Scurto, United Cable, East Lansing, Michigan

Joe Tanzarello, Production Manager, Coaxial Communications, Columbus, Ohio

Don Tennant, Don Tennant Agency, Chicago, Illinois

George Thorry, Continental Cable, Lansing, Michigan

Dotty Trommater of the Messner Agency, Grand Rapids, Michigan

John Walkmeyer, Vice-President, ATC

Johanna M. White, BancOne of Ohio

Chapter 1

Why Cable?

Cable has the unique ability to deliver *audiences of opportunity.*

GEOGRAPHICALLY DEFINED AUDIENCE

Cable can usually offer the precise geographic territory the advertiser wants. Suburban cable systems serve a single community, or several communities that are contiguous. These cable system territories generally conform to the trading areas of retail advertisers. Several different cable systems may serve a large urban area, with each system again defined in terms of areas that look very much like the retail trading areas. For example, Chicago, New York, and Philadelphia are each served by four or more cable systems. In these and other cities, the advertiser can pick, from among available cable systems, those that best suit a particular marketing strategy.

Consider, for instance, the possibility of segmenting markets, through cable, in the Detroit metropolitan area. About ten different cable companies serve a total of nearly 70 communities. The marketer interested in the upscale demographics of people in the western suburbs of Birmingham, Franklin, Beverly Hills, and Bingham Farms can make a single buy through Booth Communications and find the target. The store with brand merchandise at discount prices might go after the primarily blue-collar enclaves of Hazel Park and Madison Heights through Continental Cable. A fashion house might sell high style to the ladies of Grosse Pointe, Grosse Pointe Woods, Grosse Pointe Farms, and Grosse Pointe Park through Group W Cable.

This kind of geographic selectivity is not possible when an advertiser buys a traditional broadcast TV station like WXYZ in Detroit. Its powerful transmitter sends signals to all of southeastern Michigan and a substantial portion of Ontario, Canada. Radio is the same. Some of the radio stations in Detroit cover most of the central United States.

Aside from zoned editions of metropolitan newspapers and weekly shoppers specifically designed for designated areas, no other medium can be as selective geographically as cable TV. For the first time, the elements of sight, sound, color, and action are, or should be, available to local advertisers on a cost-efficient basis.

DEMOGRAPHICALLY OR INTEREST-DEFINED AUDIENCE

In the same way, cable advertisers can pick their desired audiences in terms of viewer interests. Many cable system territories are fairly homogeneous in income and purchasing behavior. More important, each cable system offers so many different programs that the audience for any one will represent a very narrowly defined group, either demographically or in terms of interests. The advertiser of investment services in the high-income neighborhood of Naples, Florida, may buy "Business Morning" on CNN. The golfing shop in Palo Alto, California, knows it will get serious golfers if it buys time on the first two rounds of a golfing tournament on the same network. The record store next door to the pro shop will not waste any of the audience if it advertises rock records on MTV, the rock music channel.

Advertisers interested in the people who are attracted to cultural affairs and the performing arts have discovered the Arts and Entertainment network. National advertisers have included the Greek National Tourist Organization, Piper-Heidsieck champagne, Prudential Bache, the University of Texas Press, and Yves St. Laurent. It is easy to think of local counterparts with an interest in the same audience for culture: a travel agency, wine merchant, stockbroker, bookstore, and high-fashion retailer.

To accomplish this kind of pinpoint selection, geographically and demographically, advertisers must be very precise about whom they want to reach. Media buyers have to expend more time and energy in matching cable systems and particular shows to *targets of opportunity*. Cable system advertising salespeople have to carry to prospects the message of what's available and how it's working. Then, as these media matchups are made, the creative advantages peculiar to cable can be fully realized.

OPEN MESSAGE FORMAT

Still another opportunity for advertisers in cable is to present messages of a character not possible in the conventional 30-second commercial formats of most broadcast television. When programming originates with the cable system, advertisements of any length are possible. This might be useful to a car dealer who seeks to introduce a whole line of new cars in the fall, or a department store that intends to present a television "Christmas Catalog" in November, or the bank introducing a bank-at-home, electronic funds transfer system that is much too complex to present in a few seconds.

Eventually it should be possible to use cable for direct television selling. The consumer can order merchandise seen on television by phone or, in some cases, through the two-way capability of cable. The purchase may be charged against a bank account or credit card. Commercials might be inserted in conventional cable networks, in the cable company's own programming, or in an electronic text catalog accessed through a videotex system.

PRICE AND POSITIONING

With cable advertising still relatively new, a particular advertiser can establish an advantage by being there first. Advertising rates are now comparatively low. In many cases, for pioneering advertisers, the price, the number of spots—including substantial bonuses—and positioning are all negotiable. It may be possible to obtain exclusive rights to particular programs and even entire channels. Advertisers can now learn to use cable at a low cost and under favorable conditions for experimentation.

ADVERTISER LEADERSHIP

In many places, cable operators are aggressively promoting the development of cable as an advertising medium. In other places, cable is not even available to advertisers. There are several reasons. If the system is small, staffing, administrative, and equipment costs are high relative to the revenue potential. Some cable operators are preoccupied with other areas, such as the development of multipay services and data communications. The advertiser, with years of experience in using television and other advertising media, may be more aware of the advertising potential of cable than the operator is!

It may therefore be necessary for the advertiser to take the lead and request advertising time from the operator. This may be relatively easy to do, even if the operator is not formally in the advertising sales business. Time for local advertisers is available on almost all of cable's satellite networks. The operator may also appreciate some advertising support for local programming efforts. If the advertising schedule is not too complex, manual insertion of the advertising is practical. If the operator is able to get into the advertising business in this way, without a sales staff and a major administration effort, the rates should be very favorable to the initial advertisers.

ADVERTISING AGENCY ATTITUDES

Advertising agencies, the specialists, should play a major role in the development of cable as an advertising medium. The agencies have taken different approaches. In a controversial move in 1981, the Ted Bates agency announced what it called the "5 percent solution." Bates was recommending that clients put 5 percent of their television budgets into cable to offset losses in the broadcast audiences. The recommended percentage has increased as broadcast television audiences have decreased.

For the most part, major agencies have stayed with the cable *networks*, where one buy represents the kind of audience numbers that are meaningful to big national advertisers. Buying spot time in local cable systems is too cumbersome for the sizes of audiences involved, although interconnection of cable systems may help resolve this problem.

Smaller, regional advertising agencies have had a somewhat different attitude toward cable. Dick Burkholder, of Burkholder Flint Associates in Columbus, Ohio, looks to cable as part of the media mix. He recommends it when cable is the most efficient means to get a particular message to a particular set of prospects. Burkholder says cable puts some fun back into media buying.

The enthusiasm expressed by Burkholder is in contrast to the frustration of Harold Bennett, media vice-president of the William R. Biggs/Gilmore Agency in Kalamazoo, Michigan, where no cable system sells advertising availabilities. Bennett explains that "in some of the most important Michigan markets, there was no way we could insert a local commercial into satellite network programming. Cable subscribers are only a portion of the total market to be reached, although they do represent the affluent, the better-educated, the higher-income people. The only way we can reach the rest of the market, on TV, is through the regular network outlets, or an independent."

But William Wiener of BBDO sees the end of big audiences on network TV. There are now too many slices of the pie.

A NEW BALLGAME

As suggested, television advertising will never be the same again. As cable fast becomes available to more and more Americans, and as most systems are upgraded to 35 or more channels, audiences will be segmented to the point where it will no longer be possible to make one or two television media buys to cover a market. For advertisers who do not have a truly "mass" market, cable programming will need to be studied to determine the best vehicles for cost-efficient advertising. For national advertisers, this means an administrative cost that is far higher than in the past, since there are so many cable systems and so many programming options. Media buying will require extensive knowledge of availabilities on cable networks and locally originated shows. The regional advertising agency with a firsthand, intimate knowledge of local availabilities and programming will become more important. The advertiser's own knowledge of market and product will be critical in perceptive media buying.

Advertisers, advertising agencies, and cable operators will have to work together to improve the efficiency of the system. At least for the moment, the administrative cost of buying and using television time will go up in relation to the cost of the time itself. This fact makes the media buying decisions all the more critical.

Chapter 2

How Cable Advertising Works

The typical cable television system offers off-air broadcast television channels, premium pay television services, satellite networks, and local origination and automated channels.* Advertising time is generally available on the satellite networks and local origination and automated channels. Approximately one-third of the cable systems in the country are not selling any advertising. These systems tend to be small (less than 4,000 subscribers) and also tend to be single-system, not multiple-system, operations.

Of the cable systems that sell advertising, general types are identified. The general types can almost be viewed as stages of development, although in some cases this may be an oversimplification.

Automated Channels

Often, the very first attempt by a cable operator to sell any commercial time is a simple message on either a crawl or other character-generated channel. The message may be enhanced by computer graphics. Typically, advertising revenue is very small, and the service of the accounts is generally the responsibility of cable system personnel with other principal responsibilities. The important characteristic of this type of system is the lack of any local insertion equipment.

An exception to the relative lack of sophistication of this stage of development is United Cable of Colorado, which serves much of the suburban area surrounding Denver. A relatively large system, offering 36 channels to 68,000 subscribers, United sells advertising on four automated data channels. The channels include a Program Guide, Classifieds, Home Sportsline News, and a Restaurant & Entertainment Guide. With a staff of four, United has advertising revenue of approximately $1 per subscriber per year selling only "digital" advertising.

Advertising of this type does not involve the operator in the expense of maintaining more traditional insertion and tape equipment.

Satellite Networks

The most common type of cable advertising service relies on the advertiser-supported satellite networks, such as ESPN and CNN.

*A glossary of cable and cable advertising terms will be found at the end of the book.

Local insertion equipment is used to take advantage of the advertising time made available by the satellite networks. Systems at this stage of development usually offer automated or data channels in addition to the satellite networks and often have some limited local origination as well.

A good example of this type of system is Manhattan Cable TV, serving New York City. A very large system, offering 32 channels to 170,000 subscribers, Manhattan sells advertising time on ten different satellite networks, two automated channels, and a part-time local origination (LO) channel. The local origination channel operates in a four-hour evening block for New York City independent producers and Meadowlands Racing. Part-time local origination channels commonly carry local sports events and serve as an excellent source of advertising time for the cable operator. Manhattan Cable, with an advertising staff of six, enjoys advertising revenue of nearly $1 per subscriber per month.

Figure 2-1 Automatic commercial insertion equipment makes possible the use of local advertising on several channels. (Channelmatic, Inc.)

Local Origination

Full-time local origination channels with their own production staffs are in the next stage of development. The amount of advertising time the cable operator can sell dramatically increases with a local origination channel that locally produces programming attractive to an audience. In addition, the cable operator has automated channels and the satellite networks, but the local origination is much more fully developed. The advertising support staff is considerably larger than the staff found in the less-developed systems and includes a larger sales staff and many more production and administrative personnel.

One of the best-known examples of this level of development is Palmer Cablevision, serving Naples and Marco Island, Florida. The system offers 29 channels to more than 32,000 subscribers and employs six full-time salespersons and more than 17 traffic, administrative, and production personnel. Palmer produces local news, telephone call-in shows, and programs on fishing, golf, tennis, cooking, money management, travel, gardening, and other subjects. Palmer grosses nearly $2 per month per subscriber, making it one of the highest advertising-revenue-producing systems in the country.

Most cable systems are not as sophisticated as Palmer Cablevision and will not have the support staff or the amount of time available for sale. The typical cable operator will offer between 30 and 35 channels and will be able to claim between 40 and 50 percent of all homes as subscribers. Among the 30 to 35 channels there will most often be four or five advertiser-supported satellite networks, one automated or data channel, and one local origination channel. Usually the local origination channel will not operate full time. It will carry community special events and programs such as local sports. Generally, the advertising sales staff will consist of three personnel, usually a manager and two sales representatives. Of course, this will vary substantially according to the size and sophistication of the system. The administrative support staff will be approximately twice as large as the sales staff but may have duties other than advertising.

Approximately half the operators will have automatic insertion equipment, with the others using manual insertion with an operator present. More than half will invest in some type of research to document the audience their system delivers. About one out of five will be an active participant in some type of advertising interconnect with adjacent operators and will obtain revenue from advertising sales through the interconnect. (More about interconnects will be discussed later in this chapter.) The typical cable operator will make slightly more than $8 per year per subscriber from advertising sales.

For purposes of more detailed description, we will divide cable advertising into the following categories: network and spot, local origination, sponsorship, automated channels, shopping services, classified, and cable radio.

NETWORK AND SPOT

As noted, the satellite-delivered cable networks accept advertising on the whole network (network advertising) and make local spots available to individual affiliates for national and local spot advertising. If Anheuser-Busch buys time from the ESPN cable network during a college football game, it is buying network advertising. If Prestone buys time in the same football game from several selected cable systems in the northern states, it is buying national spot advertising. When a local Ford dealer buys time in the football game from one or two cable systems in the trading area of the dealership, it is buying local spot.

The variety of satellite network programming now available for network, national, and local spot is demonstrated by the following list of cable networks.

ADVERTISER-SUPPORTED CABLE NETWORKS

Name	Programming	Number of Affiliates*	National Subscribing Households*
ARTS and Entertainment	Visual and performing arts	2,000	12,500,000
BET (Black Entertainment TV)	General, black-oriented	432	9,033,719
CBN Cable Network	Family entertainment, some inspirational religious	5,551	28,017,000
CNN (Cable News Network)	24-hour news and news features	7,780	32,262,000

Name	Programming	Number of Affiliates*	National Subscribing Households*
CNN Headline News	24-hour hard news, cycled every 30 minutes	2,403	16,284,000
Country Music TV	Country music videos	314	4,900,000
ESPN (Entertainment and Sports Network)	Sports events, sports and business news	9,900	35,700,000
FNN (Financial News Network)	Business news, stock and commodity quotes, sports scores	1,050	18,100,000
Lifetime	Health, fitness, science, women's interests	2,400	24,900,000
MSN—The Information Channel	General information, consumer-oriented	510	10,620,000
MTV (Music TV)	Rock music videos	3,100	26,200,000
National Jewish TV	Jewish	247	5,569,000
Nickelodean	Preschool through teens	3,800	24,900,000
SIN (Spanish International Network)	Spanish language entertainment and information	348	29,405,100
SPN (Satellite Programming Network)	General entertainment and information	525	11,800,000
TNN (The Nashville Network)	Country-oriented entertainment	2,250	22,160,000
USA Network	Sports, women's interests, children, entertainment series	4,200	29,700,000
The Weather Channel	24-hour weather news with local information	1,630	17,000,000
WGN, Chicago	General (no local avails)	6,515	15,714,000
WOR, New York	General (no local avails)	1,378	5,231,220
WPIX, New York	General (no local avails)	152	1,385,000
WTBS, Atlanta	General (no local avails)	8,656	33,111,000

*Affiliate and subscriber counts 5/84. Source: *Cablevision,* August 3, 1984, p. 46. Subscribing households represents cable households with access to the network, not necessarily the households that make use of the networks.

USA NETWORK

CONTACT: MaryLou Brown
(201) 445-8550 X241

BEST OF USA #1
(1983 Suntory Cup Tennis Finals--Borg vs. Connors)

COMMERCIAL FORMAT #55

1982-1983 SEASON

PROGRAM LENGTH: 2 HOURS FIRST AIR DATE: JULY 7, 1983

(N) NATIONAL (L) LOCAL

OPEN/BILLBOARDS/INTRODUCTION

Commercial Break #1	(N) 2:00

START OF MATCH...

Commercial Break #2	(N) :10/(L) 1:00
Commercial Break #3	(N) 1:30
Commercial Break #4	(N) 1:30
Commercial Break #5	(N) 1:30
Commercial Break #6	(N) 1:00/(L) 1:00
Commercial Break #7	(N) 1:30
Commercial Break #8	(N) 1:30
Commercial Break #9	(N) :10/(L) 1:00
Commercial Break #10	(N) 1:30
Commercial Break #11	(N) 1:30

END OF MATCH

Commercial Break #12	(N) 1:30/(L) 1:00

CLOSING COMMENTS/BILLBOARDS/CREDITS

###

208 HARRISTOWN ROAD, GLEN ROCK, NEW JERSEY 07452, TELEPHONE (201) 445-8550

Figure 2-2 Commercial lineup for a tennis match. Commercial breaks designated (L) are local availabilities.

By and large, cable systems select those satellite channels that make the best sense for the communities they serve. Since cable systems may carry anywhere from twelve to more than 100 channels, there is considerable variation in the number of satellite channels offered. The high-capacity systems are likely to have all, or almost all, the channels just listed.

The advertiser-supported cable networks will permit sponsorship of entire programs or sell ad spots within programs. These commercials are inserted by the network and go with the program to all cable system affiliates. Each of these networks also makes available a designated number of minutes within most hours for the cable affiliate to use locally. These are called *local availabilities*, or, for short, *avails*. Retailers and other advertisers can buy the local audience created by these national networks by contracting for the local availabilities. Regional or national advertisers may also buy local avails over several cable systems, rather than the whole network, to reach a specific geographically or demographically defined market. This task is made administratively easier by the interconnection of cable systems regionally.

A special category of cable channel serves national advertisers. These are called *superstations*. The superstations are actually broadcast stations licensed to a particular city that are picked up by satellite for distribution to cable systems nationally. The first of these, WTBS in Atlanta, is owned by cable pioneer Ted Turner. The other three superstations are WOR and WPIX in New York and WGN in Chicago. The superstations are available to local advertisers only in the markets in which they originate.

Archway, the nation's fourth largest cookie baker, uses cable networks, superstations, and national spot. Don Keaton, vice-president and marketing director of Archway, operates an in-house advertising agency. He points out, "We're not a giant advertiser like our major competitors, National Biscuit and Keebler, so we have to try to buy more gross rating points for less money."

Archway got more than it had bargained for when Keaton, as an experiment, bought a schedule of WTBS and CNN. Each year, just before Thanksgiving, Archway brings out a Holiday Pack of attractive decorated cookies. The commercial used was based on the fact that all the company's advertising had, for years, been primarily directed at adult women, the prime purchasers of cookies. So when a jolly Santa

Figure 2-3 Ted Turner, of Turner Broadcasting (WTBS, CNN, CNN Headline News). Originator of the superstation and all-news television and the father of advertiser-supported cable networks. (Turner Broadcasting System)

Claus suggested a plate of holiday cookies by Archway next to the chimney on Christmas Eve, it was intended as a bit of a spoof. And there was a twinkle in St. Nick's eye when he closed the commercial by saying, "So be sure that those cookies are waiting for me . . . because I'm keeping a list!"

To an adult, perhaps, it was a touch of humor. But to children, it meant something else. The message did reach children through some of the programming included in the WTBS buy—and the children reached their mothers. Tearfully, they insisted that Mom had to buy Archway cookies or Santa wouldn't come to their houses. Across the country, mothers reacted. By telephone and letter, they irately protested this seeming pressure on their youngsters. The commercial was pulled and the ending changed. Letters and free Holiday Packs went out to the complainants.

Now, as a supplement to Keaton's regular network buy, cable is very much a part of his media planning. Two aspects that he particularly likes are the make-goods and the relatively low cost.

"WTBS promised us so many gross rating points," he says, "and we didn't quite make it. So on my next schedule, we'll get makeups as

Figure 2-4 The fast pace and on-beat cuts typical of MTV programming are picked up and used in this Archway commercial designed for use on the music channel. The storyboard only suggests the action and music that will support the lyrics. (Courtesy: Archway Cookie Co.)

well as whatever we buy." In the future, Keaton sees definite possibilities for using local cable availabilities in conjunction with the nine regional bakeries where "the big cookies with the small-town taste" are produced.

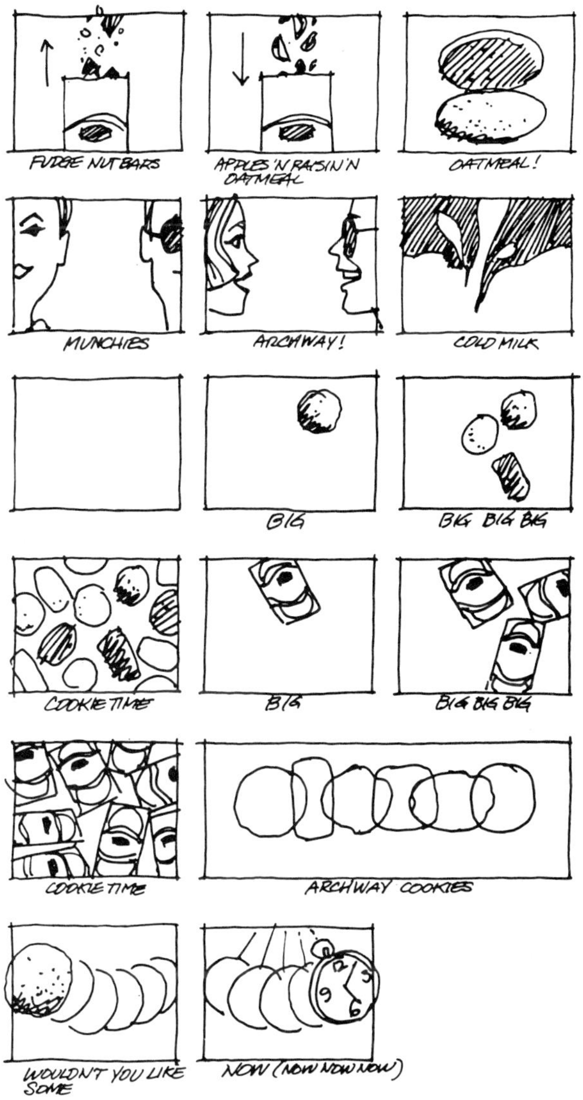

Figure 2-4 (Continued)

"Because of our budget and marketing needs," he concludes, "network and spot TV are still primary with us. We've got to know that we're reaching the kind of big numbers they produce. But as cable becomes able to generate and document numbers, it will be more and more of a factor." Recently, when marketing information indicated an opportunity to increase sales to the 18–34 age group,

Keaton moved into cable again. A special musical commercial was prepared and scheduled on MTV.

LOCAL ORIGINATION

We have so far discussed cable satellite networks and spot advertising on these networks. Most larger cable systems also have their own television production studios where they originate local news and other programming. They may also have remote units capable of producing local sports and special community events. This programming is referred to as local origination, or simply LO. Advertisers may be able to buy 10-second, 30-second, 60-second, or longer commercials on locally originated shows. It is here that the local advertiser has the most freedom. When the right audience can be reached and the commercial format is appropriate, such shows can be a good advertising buy.

A few cable systems also program one or more local origination channels with syndicated off-network reruns and movies. Such channels may look very much like independent broadcast television stations, with reruns of "M*A*S*H," "Happy Days," and "Leave It To Beaver," as well as some elderly films that have already had their broadcast runs. Since cable operators cannot pay as much for syndicated programming as the broadcast stations do, they get the least desirable and oldest of available syndicated programming for the market. Nonetheless, these programs may be an attractive viewing alternative for some people, and a good buy for advertisers if priced right. Gill Cable of San Jose, California, offers a special channel called the Classic Movie Channel. An avid movie buff can view a parade of old films. Advertisers interested in the audience can buy access to film lovers from a potential of 350,000 subscribers.

SPONSORSHIP

A national advertiser can buy a complete program from a cable network, thus sponsoring the program. Or the national advertiser may, under agreement with the network, put together a complete program package, including the commercials, to run on the network. This, of course, is the ultimate sponsorship.

Cable companies with local origination channels would also welcome programs designed for specific advertisers that would benefit from close identification with the show.

A special kind of local origination is called "access programming." The FCC used to require access programming or access channels from all large cable systems. This is no longer the case, but many local franchises still require the access channels. Public access is available to anyone on a first-come, first-serve basis, except for advertisers with commercial messages. Other access channels may be devoted to education and government. *Underwriting* may be permitted on such access channels in some franchise territories. Underwriting is the public-spirited support of programming for which the funding source is identified. The underwriting may even carry a promotional message, called "enhanced" underwriting. It might say, "This program was brought to you in part by the First National Bank of Middleville, Middleville's first full service bank, with branches in 25 locations." The concept of underwriting, of course, comes from public broadcasting. If permitted on certain access channels and available in local origination, it is a good opportunity for institutional advertisers.

AUTOMATED CHANNELS

Almost all cable systems carry a number of text channels, such as the Dow-Jones stock prices, Reuters News Service, a shopping information channel, the weather, and so on. These text services (called "data," "automated," or "alphanumeric" channels) may be divided by color into two or more horizontal sections, with one of the sections available for text advertising. Since these channels are used for only brief periods of time by any cable subscriber, the advertiser can buy long periods of time, sometimes exclusively, on the channels. It is easy to change the text, so the advertiser has an opportunity to keep customers updated on specials and promotions.

A good example is the business liquidator who bought the crawl on the time and weather channel in Rockford, Illinois. The merchandise available for sale depended on the businesses liquidated. The crawl could keep potential customers informed of what was available at bargain prices at any given time.

SHOPPING SERVICES

Cable systems may offer several types of shopping channels. A channel may be exclusively devoted to advertising, with advertisers presenting messages and programs that are entirely promotional. It has been reported in studies by the J. Walter Thompson Advertising Agency and Adams-Russell Cable System experiment in Peabody, Massachusetts, that there is a surprising consumer interest in such an all-commercial channel.

Other shopping information may be organized as an electronic catalog, whereby the subscriber to the cable system or to this particular service accesses different categories of merchandise through a videotext system. This is a two-way cable service whose subscribers can actually order the merchandise over the cable system.

CLASSIFIED

Cable offers classified ads as well as more conventional television advertising. In fact, cable adds new dimensions to classified. Some cable classifieds are part of the videotext or teletext system in which the user calls up any product or service category desired, using the code numbers listed in a master index. Since only a few cable systems have this capability, the alternative is to present a continuous, running, alphanumeric classified channel. Here, the seeker of a particular item has to wait until that category comes up in the cycle. This is much less convenient, but those who are fascinated by classified ads generally, or are just looking for a good buy, whatever the item, will be satisfied and maybe even entertained.

To be innovative, cable has added a new dimension to classified advertising, the visual. The ATC cable system in Fayetteville, North Carolina, takes visual classifieds. The user can go to a drive-in window where a picture will be taken of the item, or any kind of visual material can be brought in with the copy. In one two-week period, the Fayetteville cable system sold 26 used cars for its classified customers.

CABLE RADIO

Cable systems may offer a variety of audio services. The cable operator may put an audio service, in which spots can be sold to advertisers, behind a text channel. Cable systems may also offer FM radio services via cable satellite networks whereby local availabilities and national spots are available. These networks can operate the same as the video networks. A considerable variety of network audio service is becoming available to cable operators. Thus, the advertiser can get access to a fairly targeted audience defined by program interests.

It is also possible for a cable system to originate radio programming. This programming can have a local character and fit a particular local need. If cable radio origination attracts audiences, of course it is of interest to local advertisers.

A case in point is LAN Cable Radio, launched in May 1982 in Lansing, Michigan. LAN has access to about 60,000 households in the Greater Lansing area through two cable systems. In Lansing, it is on video channel 29 (alphanumeric text) and the FM audio band in stereo. In East Lansing and Meridian Township, it is carried as stereo on the FM band. The music is African-American cultural, filling a

Figure 2-5 African-American music featured on advertiser-supported cable radio station LAN. (LAN Radio, Lansing, Michigan)

void in the Lansing market. Information segments at five minutes to the hour are a series of public service announcements (PSAs) on what's going on in town.

Advertisers have included the U.S. Army, Whistle Stop take-out, McDonald's, political candidates, real estate agents, Jay Vincent's Records and Tapes, KZ Men's Wear, Ebony Production Concerts, Grandma's Kitchen, Spencer-Brook Fashion Design, Capital Cadillac, and barber and beauty shops. Many of the advertisers are simply unable to afford any other advertising medium. Others are looking for that segment of the community attracted to the African-American orientation of the station. The effect of advertising on LAN Cable Radio is most obvious to new businesses that have used the station to help them get started.

LAN Cable Radio recommends a minimum four-week contract of twelve 30-second spots per week, at a total cost of $200. The cable station is liberal with bonuses in unsold time to make sure there is a response for the advertiser.

These cable radio, or audio, services may be most successful in smaller towns, such as Lansing, with a limited number of radio formats. Listeners (and advertisers) will not be forced to relate to the usual middle-of-the-road format, but can have available to them specialized stations for their special interests.

THE CHALLENGE

Technologically, cable offers clarity of picture and sound. Demographically, it offers ever-increasing reach and choice, and a continually growing body of viewers. For the advertiser, the overriding question is now how to best use cable to reach prospective customers with maximum efficiency.

Chapter 3

Some Cable-Only Advertising Opportunities

If you are still thinking of cable as just another form of TV, consider what happens now.

DATA CHANNELS

This development, now part of the programming mix on most major cable systems, turns your TV screen into something you *read* rather than *watch* and *hear*.

For up-to-the-minute news, weather, sports, stock reports, community activities, or program guides, turn to the indicated data channel. Information will flash across the screen, and be repeated and updated periodically, usually on a 24-hours-a-day basis. This material may be interspersed with appropriate advertising, in either text or graphic form.

Tulsa Cable Television, a 35-channel system, has worked hard at making this kind of advertising useful to local merchants and profitable to the system. Clients include shoe stores, beauty salons, insulation firms, audio equipment stores, appliance dealers, and a vocational training school. Available to them are five channel choices:

1. *Weather Channel.* A single-line crawl of 256 characters can be inserted into each local report (approximately every three to five minutes). A single advertiser could preempt the advertising portion of the report for the whole month at $1,000.
2. *Program Guide Channel.* Three pages (14 lines, 32 characters each) of advertising every half hour, available for $1,000 per month per page. Limited to three advertisers.
3. *Sports Information Channel.* Available to one advertiser for $300 per month.
4. *Classified Ads Channel.* Fourteen lines cost $7 per day; three lines are $2 per day.
5. *Stock Market Quotations Channel.* One advertiser can buy a five-line, 32-character crawl for $450 a month.
   (The rates quoted are subject to change.)

As of a 1983 report, all advertisers on these channels were local. One, the vocational training school, reported that it had used every conceivable medium, but nothing worked as well as the system's Program Guide Channel.

United Cable of Colorado, serving the Denver metropolitan area, began operation of its data channels in November 1982. In its first six months, it generated $25,000 in sales. Each of the channels seemed to be particularly appropriate to a certain group of advertisers:

> *Program Guide:* carpet store, child-care center, travel agency, car dealership
>
> *Classifieds:* small businesses, discount food store, car dealers
>
> *Home Sportsline News:* video and record stores
>
> *Restaurant and Entertainment Guide:* catering houses, restaurants

The Program Guide offers four advertisers the chance to have full-page ads of 25 words run many times daily for $1,000 a month.

Classifieds run 144 times a day, and copy can be changed daily.

Full-page ads are also available on both the Home Sportsline News show and the Restaurant and Entertainment Guide.

The *Herald Leader Telepress* of Lexington, Kentucky, was set up by the Knight-Ridder newspaper chain for research and development and to gather data regarding viewership trends. Telepress news is programmed by the newspaper partner of the cable operation, the Lexington *Herald Leader.* The newspaper's ad staff, consisting of 12 full-time salespeople, sells the data channels along with regular space sales. There are three such channels.

The *Telepress News* is a 24-hour headline service with 20-minute continuous loops of news updates, cable and theatrical movie reviews on weekends, and a continuous weather crawl with day, date, time, and temperature. During one survey period, viewing of the news channel in households where it was available climbed from 25.3 percent to 35.1 percent. Phone surveys indicated that 23.3 percent of respondents found the news channel the best source of immediate news and information.

Advertisers can buy into this audience in four ways, ranging from a simple 320-character crawl that runs six times an hour for $400 a month, to sophisticated computer animation. The computer animation interrupts the news and commands greater attention. It is priced at one per hour for $150 a week, plus a $75-per-hour fee for the artist.

Telepress Sports and Stock Reviews supplies scores, standings, racing results, features, and A.P.'s Most Active Stocks, all on 30-minute continuous loops. One stockbroker advertiser reported selling products featured on financial news updates and stock crawls.

Telepress Bulletin Board and Classified Channel offers 24 hours of product and service listings intermixed with community events and infomercials. Classifieds cost $1 a day and run at least 24 times. Infomercials (primarily sold to car dealers) can run up to five minutes in length and cost $100 for five daily showings.

With these three data channels in operation, the system generated $120,000 in sales for 1983. The biggest problem facing salespeople is the inability of advertisers to relate to this form of advertising. Finding a special place for this type of advertising in the marketing mix isn't easy for advertisers accustomed to traditional media.

These examples in the Tulsa, Denver, and Lexington markets indicate the untapped possibilities for local advertisers to reach important audiences in an effective if unorthodox way.

THE CABLESHOP

The idea of a TV channel that offers nothing but commercials was certainly impractical before cable. With the advent of many available channels, there was still considerable doubt as to the attractiveness of a "TV store," the sole purpose of which was to sell merchandise and services. But in March 1982, the Peabody Pilot Test, a major experiment in advertising, was launched in Peabody, Massachusetts. It was co-sponsored by Adams-Russell, a system with 7,000 subscribers, and the giant advertising agency, J. Walter Thompson. After an experimental phase of nine months in the Boston area, six major market areas were added: Los Angeles, through Rogers Cablesystems and *Times-Mirror;* Detroit, through Maclean Hunter and United Cable; Chicago, with Cablenet; Tucson, with Cox Cable; Erie, Pennsylvania, through ATC; and Metro New York, with Maclean Hunter. This combination made it possible to reach 243,500 subscribers with messages from 17 national advertisers, plus regional and local sponsors. Market research based on 14 national brands using Cableshop indicated a 19 percent increase in brand awareness, a 30 percent increase in intent to buy, and a 37 percent increase in recent purchase.

More significant than these figures are the attitudes indicated by viewers of the long-form commercials, called infomercials, that appeared on Cableshop. They considered them "interesting, factual presentations . . . learning experiences that help them make better purchase decisions."

Programming for this 24-hour video shopping service was a combination of long, informative commercial messages prepared by advertisers and how-to consumer videos with useful information on the purchase of everything from cantaloupes to computers.

Cableshop planned to go national via satellite and become available in many additional cities. Its announced intention was to make time available for *local* infomercials. However, lack of sufficient revenue led to a closing down of Cableshop in late 1985.

THE CABLE SAVER CLUB

Individual cable system operators are developing new ways to attract local businesses to the medium.

"In addition to attractive programming, we've developed other ways of making ourselves increasingly valuable to advertisers," says Steve McVoy of Coaxial Communication in Columbus, Ohio. He is talking about a wallet-sized green, yellow, and white card that signifies membership in the Cable Saver Club, an idea copyrighted by Coaxial Communications. Cardholders are entitled to discounts and specials with area merchants.

Both McVoy and his ad director, Dave Gettys, are enthusiastic about this innovative approach to the marketing possibilities of their business. "First," says McVoy, "it gives the person who is considering subscribing to cable another reason to join the system. It can help sell installations."

"Then," adds the advertising director, "it offers local merchants a reason to use us as a marketing tool."

Some 50 retailers in the area served by Cable Channel 6, Columbus, are already enrolled as participants in the Cable Saver Club. This entitles them to point-of-purchase materials identifying them to

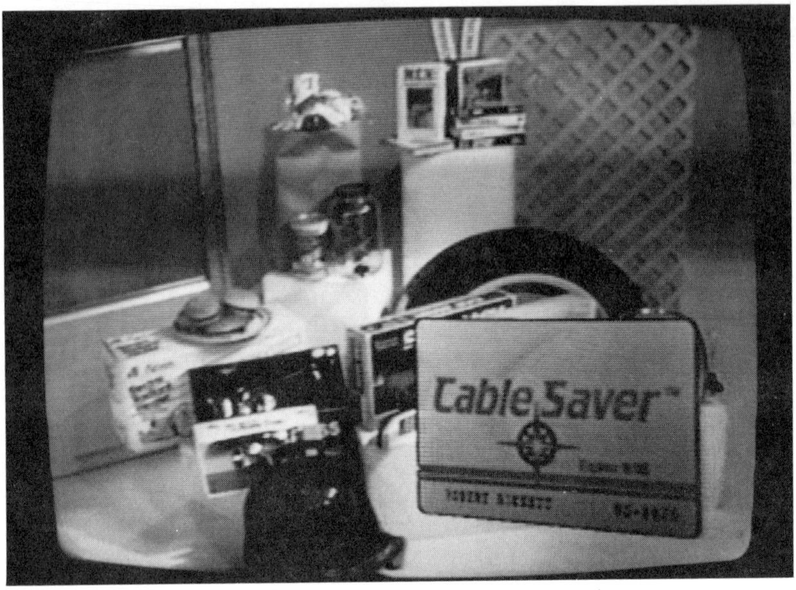

Figure 3-1 Display of discounted merchandise available only to cable subscribers. (Coaxial Communications, Columbus, Ohio)

Channel 6 subscribers as places where preferential pricing can be expected. Exactly what form such pricing takes is explained daily on Channel 6. Text messages give this sort of information:

> Ajax Brakes—15% Off on the Parts Installed Only
>
> The Bake Shoppe—Buy 1 Dozen Danish Sweet Rolls or Donuts and Get ½ Dozen of Either Free
>
> General Tire Sales Company—20% Off on New Tires, 10% Off Automotive Services—Alignment, Brakes, Shocks, and Lube
>
> Jean Scene—Special Anniversary Discount from the Jean Scene Coming Later this Month, Watch for It
>
> Kinder Care—No Registration Fee when Enrolling Child (Save $20.00)
>
> Neil Pharmacy—10% Off on All Prescriptions and 10% Off on Medical Equipment (Rental or Purchase). Plus a Free Gift Worth Up to $5.00 When You Show Your Cable Saver Card

> Willard's Restaurant—10% Off Any Dinner—Tuesday, Wednesday, and Thursday Only After 4:00 P.M.

Pizza parlors, car washes, record shops, rent-a-cars, flower shops, weight-loss programs, hardware stores, hair salons, furniture stores, paint centers—all are represented when the text rolls on Channel 6. Merchants involved in the project acknowledge its success. Some of them have supplemented the listings with commercials on other available networks. All of them have a new awareness of cable as a way of reaching prospects. The Cable Saver Club is one of those ideas whose implementation takes a lot of energy, work, money, and conviction. It's like removing all the surface soil before finding a rich vein of precious metal. However, once the *productivity* of the medium is proven, the need for ratings, viewership counts, and the like disappears. In simple terms, the Cable Saver Club is a free window into 56,000 homes, and, like most showcases, it attracts customers.

DIRECT RESPONSE

Some idea of the possibilities that cable offers for direct selling can be gathered from available data relating to its performance for direct-response advertisers. Here, cable acts as a middleman, taking orders on toll-free lines and collecting revenue based on the number of calls that come in in response to the commercial. The following are examples:

> WTBS accumulated sales of more than 100,000 albums for a record offer that lasted six months.
>
> CBN Cable receives about 80,000 product orders or inquiries on its toll-free lines each month.
>
> Cable Health Network (now Lifetime) generated some 20,000 responses for *American Health Magazine*.

In direct-response advertising, the advertiser does not have much to lose, since the cost of the advertising is usually a percentage of the sales or a fixed price per lead.

THE CATALOG

Certainly the presentation of a complete catalog through the medium of television would have been impractical before cable. But in 1983, Richard Thalheimer, the brilliant marketer who originated the *Sharper Image* catalog, introduced to the cable audience live demonstrations and previews of catalog products (see Figure 3-2).

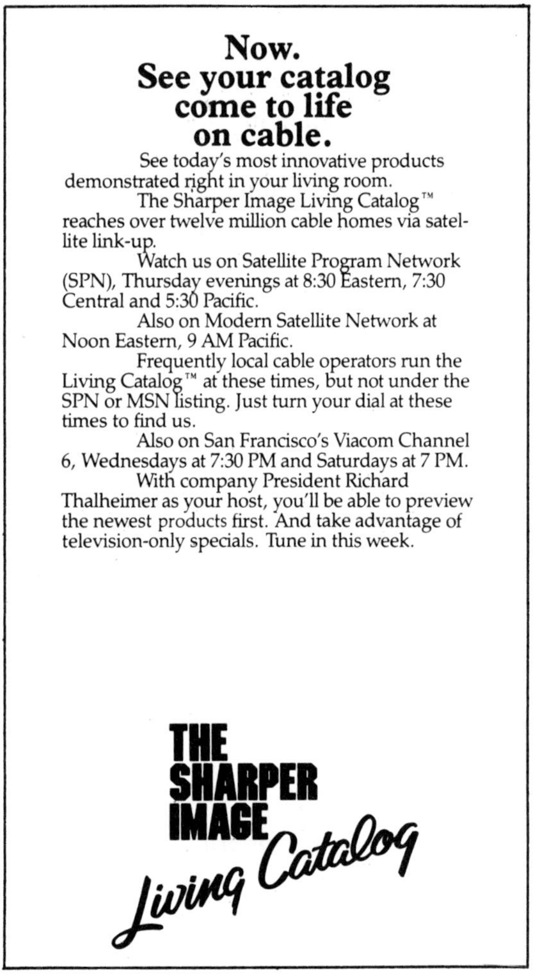

Figure 3-2 This insert in the Spring 1983 Sharper Image printed catalog is an invitation to see products in action. (Courtesy: Sharper Image Catalog)

With the Living Catalog as a registered trademark, the San Francisco company used merchandise as entertainment and attracted viewers through what was, essentially, one long commercial.

CABLE PROGRAM GUIDES

Another unique advertising opportunity for advertisers in cable is the printed program guide circulated by many cable systems. It is increasingly necessary for the cable operator to supply a guide to programming, since there are so many channels. The subscriber who is not aware of the array of programming available is not fully utilizing the service and is less likely to be retained as a subscriber. Ideally, the cable programming guide will be system-specific. That is, the guide will list every single channel carried by the system, with every program listed and described. The channels will be listed by the number on the converter assigned to the channel. The local advertiser can add mileage to video advertising by buying space in the guide promoting the programs sponsored and the products advertised.

Space in the guide may be sold by the cable operator or directly by the guide publisher, usually an independent organization.

CONCLUSION

Data channels, shopping channels, Cable Saver cards, televised catalogs—all offer new approaches by the seller to the buyer. None were really feasible until TV set owners had many choices as to what would appear on their home screens at any given time. It's not clear yet what effect this will have on either entertainment or advertising. What is clear is that cable is a distinct and viable medium gaining an ever-increasing share of dollars expended on commercial communication. In addition to these new possibilities, some of the values of television advertising have been restored, particularly sponsorship.

Chapter 4

Opportunities for Sponsorship

Sponsorship of attractive, well-liked programs providing a hospitable environment for commercial messages represents an advertising value that may translate into purchase and loyalty. Such sponsorship opportunities are abundantly available on cable.

In earlier times, when radio was the major broadcast medium, ownership of shows by single advertisers and identification of stars with particular products was considered highly desirable. From the Voice of Firestone to the A & P Gypsies and the Ipana Troubadors, product names found their way into program listings. For years J-E-L-L-O and Jack Benny were synonomous, as were Texaco and Ed Wynn, the Fire Chief.

Television, too, with its GE Theatre, Dinah Shore Chevy Show, United States Steel Hour, Colgate Comedy Hour, and others, allowed the advertiser to be the owner. Compatibility between program and product was a reasonable objective from the 1950s through the 1970s.

However, along with audiences of unparalleled size, TV began to deliver costs that made sponsorship unaffordable. The 30-second spot became the commercial norm. Rating points, rather than show content, decided where spots would be placed.

Now cable, with its specialized programming, smaller audiences, and lower costs, has made sponsorship a viable proposition once again.

LESSONS FROM THE PAST

With that in mind, it is constructive to review a few sponsorship case histories.

From the late 1930s to the early 1950s, the Union Electric Company of Missouri sponsored a locally produced program on St. Louis radio stations. A dramatization of local history, "The Land We Live In" enjoyed excellent ratings and played to live audiences in the company's auditorium. Many of the programs had direct tie-ins to local charities, industries, and organizations. With the advent of TV, ratings for this radio show began to slip. An attempt was made to convert to the new medium, but costs were prohibitive and the program was reluctantly dropped. In the ensuing years and up to the present, the company continues to get letters from businesses, indi-

Figure 4-1 The General Electric Theatre, with its carefully selected dramatic fare, gives substance to the company's corporate statement, "We bring good things to life." (Courtesy: General Electric Co.)

viduals, schools, and other organizations asking for information contained in the scripts.

J. Wesley McAfee, former chairman of the utility, considered "The Land We Live In" the best public relations mechanism in the company's history and would never have considered anything except sole sponsorship.

For institutions concerned with public regard, some type of similar programming could be created and placed on a cable system with facilities for local origination. Tie-in promotions with organizations featured in programming could increase viewership and win friends for the sponsor. Within the next few years, an entire metropolitan area or state may be reached through interconnects. That will make it possible for regional advertisers to reach selected audiences throughout the areas they serve.

Full sponsorship of special entertainment events was another way of winning friends for corporations. The two Fred Astaire specials sponsored by Chrysler were TV landmarks in 1958 and 1959. Some idea of the public response can be garnered from comments by critics:

> "An hour that will rank among the great contemporary theatre nights in any media." ——Jack Gould, *New York Times*.
>
> "Commercials of Chrysler's 'Can Do' line of five cars by the Leo Burnett Company were practically an auto show in themselves, by far the best of any to date." ——*Variety*.

When one thinks back to Gillette's identification with boxing, Texaco's longtime sponsorship of opera on radio, and Hallmark's consistent presentation of fine drama, it's apparent that each company sought an identity with a particular audience. With cable's manifold programming choices, the selection process is becoming far more exact.

SPONSORSHIP RETURNS ON CABLE

In 1981, Continental Cablevision of Concord, New Hampshire, embarked on an ambitious program of covering local sports on its local origination channel. For from three-and-a-half to four-and-a-half hours a week, subscribers could watch New Hampshire college basketball, high school championship games, and even the Pop Warner Football League games involving boys between the ages of 8 and 14. There was a lot of related promotion, including a Capitol City Trophy for the school with the best all-around record in interscholastic sports. Four dozen local sponsors bought the four minutes of ad spots available per half hour. Of most interest to advertisers should be the fact that 42 percent of the system's subscribers could identify the sponsors of local sports coverage. It would be interesting to see that fact measured against the recall of advertisers on the regular networks during the same time periods.

Local origination is affordable. Spots have been available on some locally originated shows for from $10 to $50 per 60 seconds. An advertiser could actually control six spots an hour, year-round, for as little as $125,000 a year. Compare that with the cost of a continuous campaign of a year's duration in any other local medium.

Burl H. Robins, creative director of Stevens, Inc. Advertising in Grand Rapids, and agency president David Vanderveen have implemented a tailored use of cable for their client, Amway, of Ada, Michigan.

"It is ideal for us," Robins explains, "because we know exactly whom we want to reach. Our audience is the distributor, or potential distributor, of Amway products. Through direct mail and newspaper advertising, we let them know where and when we'll be telling our story. We suggest that our distributors have friends and neighbors in to see the show. In a sense, we use cable like some companies use closed-circuit television."

Robins makes the point with a small newspaper ad (Figure 4-2) that was run in the *Dallas Morning News, Fort Worth Star Telegram, Pittsburg Press, Atlanta Constitution, Tampa Tribune, St. Petersburg Times, Ontario Report, Pomona Progress Bulletin, Phoenix Republic, Kansas City Star, Buffalo News, Portland Oregonian, New Orleans Times Picayune, Indianapolis Star,* and *Columbus Dispatch Citizen Journal.*

Program content is a 27-minute film, already in the can, originally designed for Amway meetings. Three commercials originally prepared for network TV are slotted in, and the 30-minute time segment is filled.

Results? Distributors indicate that the effort is extremely useful in recruitment and ask for more of the same.

Figure 4-2 Through the use of cable TV in selected markets, Amway talks to people about the opportunities the company offers. (Courtesy: Stevens, Inc. Advertising)

The idea of "narrowcasting" had particular application for Amway in early 1983. A problem in public and company relations particularly affected the Detroit-Windsor area. Using all available cable systems in the area, Robins told the Amway side of the controversial issue. A special 30-minute show was written and produced. The first half concentrated on the excellence of the company's products. It gave Jay VanAndel and Richard DeVos, the company principals, a platform from which to speak. Viewers had a look at the size and scope of Amway's facilities. The concluding part of the show emphasized recruitment, and a 90-second infomercial, with Bob Hope, was slotted in between sections. The cost of the program, when compared to a similar effort on network or independent TV stations, was minimal, says Robins.

BREAKING COMMERCIAL STEREOTYPES

"That, in fact," Robins declares, "is what makes cable so attractive. There's a chance to experiment, develop, try things, without tremendous risk. It's an ideal place to test ideas as well as reach select audiences."

In any case, he sees the low cost of both time and production as a great chance to break commercial stereotypes and try new ways of reaching particular audiences with the Amway story.

"Cable also makes a lot of sense for the local advertiser looking for an audience with upscale demographics," he says. "If I sold expensive furniture, new cars, jewelry, I would certainly think about cable as a carrier of my messages."

Robins acknowledges that some of the problem is with the cable system operators. "They often seem to lack the drive, the people, the attitude it will take to realize cable's potential. Maybe," he speculates, "they just don't comprehend what they've got."

"I can imagine," Robins considers, "that large Amway distributors might be interested in local programming opportunities, but it is up to the systems' operators to make those opportunities available and attractive."

His sentiments are echoed by Kevin Patrick Smith, the special projects manager of NYT Cable TV, Audubon, New Jersey. Smith believes that complete marketing strategies, including merchandising and promotion as well as advertising, must be developed for clients. They must be able to directly attribute sales to cable. Shows developed for local origination are one avenue to this objective. There is, for instance, the "Cable Rock Around Show," which includes fashion tips, rock rap, movie reviews, and special rock features in its format. It grosses about $750 in ad revenue per show.

To quote the Cable Television Advertising Bureau:

> NYT Cable has concluded that advertising sales must be integrated with sponsorship opportunities and merchandising promotion efforts. It not only helps to convince advertisers that cable is extremely effec-

tive, it also helps the system build and maintain viewership to their channels. NYT Cable has generated over $100,000 in billings, which certainly helps to prove their point.

Palmer Cablevision of Naples, Florida, has developed local shows on fishing, golf, tennis, cooking, travel, and money management in order to provide vehicles for advertisers who see advertising advantage in sponsorship of such shows.

Coaxial Communications in Columbus, Ohio, presents a prototypical case of the active advertiser-supported local origination programmer. According to Joe Tanzarello, production manager, this system's local origination has three main categories:

The first, high school sports, is the most successful. There are twelve schools in the cable franchise. One football game and two basketball games are televised each week of the season through the championship games of their respective seasons. Each of the schools is represented at least once on TV. The games are played on a tape-delayed basis right after the event and then twice more on succeeding evenings. Sponsors include pizza parlors (where kids gather to watch the games), a local technical school interested in reaching high school people, and Burger King, which wants to be recognized for its involvement in community activities.

The second type of sponsored local origination is Annual Special Events—the Pickering Labor Day parade, the White Hall Fourth of July parade, the Reynoldsburg Continental Amateur Baseball Association 12-and-Under Tournament, the Budweiser Lite Tennis Tournament, and so on. This is a wholesome slice of American life, and Columbus merchants are anxious to be a part of it.

The third category of sponsored origination is called the Series. These are several programs put together as a series package and usually repeated one or more times. Examples are the "Jazzercise" series, six one-hour shows sponsored by a retailer called Dancewear and Things; and "Kaleidoscope," a series of twelve one-hour shows each featuring a different local band and comedian. The latter were taped on location at Mr. Larry's East (a bar) and the Hyatt Regency Center Club. Sponsors were an independent grocery store (how often are they on TV?), a hair-care salon, and a sandwich shop.

Other cable systems have developed interesting program concepts. According to *Electronic Media* magazine, five Southern California operators run "Hollywood Weekly," a nightly neighborhood guide to local movies. Movie companies pay to be listed. A "TV Automart" in Los Angeles features private-party automobile sellers. A Phoenix "Home Channel" has 24 hours of residential real estate information daily. *Electronic Media* quotes the programming director of the Southwest Region of Group W Cable, Ardie Ivie: "We're not trying to duplicate or compete with entertainment services we carry on our other channels, but we believe that informational needs, whether commercial or not, can be developed on a local level, and we hope that the business community will be a major supporter of these services."

THE AGENCY'S ROLE

There was a time when advertising agencies considered the creation of appropriate programming for their clients part of their job. All-time leader in the field was the Blackett-Sample-Hummert agency of Chicago. In the 1930s, it had as many as 67 shows a week on the networks. Most major agencies had very active program departments with the responsibility for developing properties that would serve the needs of clients. The tone, content, and stars of such properties were given a great deal of attention by people like Phillip L. McHugh, broadcast vice-president at Detroit's Campbell-Ewald Agency in the 1950s and 1960s.

It was McHugh who helped devise, for Delco batteries, the "High Adventure Show" with Lowell Thomas. It quickly won respectable ratings, but, more important, it provided a showcase for remarkable, integrated commercials that dramatized the dependability of the product in exotic settings around the world. Research indicated amazingly high commercial recall and retention of the ideas about the product that were written into the advertising. Critics and columnists were fascinated and pleased with the absolute cohesiveness of the package.

Ralph Porter, a top writer on broadcast subjects in the 1950s, said, "A documentary smack of reality and immediacy flavors these commercials. It is this dynamic approach, so sorely needed in television selling, that justifies the integration experiment of 'High Adventure.'"

The word hadn't yet been coined, but he was discussing what would eventually be called infomercials—now a buzzword in cable advertising.

McHugh was also deeply involved in the creation of special programs for the General Motors Corporation. Working with Gail Smith, who was to become general director of advertising for the automaker, Campbell-Ewald developed the first appearance on TV by Danny Kaye, whose film career had made him one of the country's favorite entertainers. Here again, the agency and the client worked out some of the prerogatives of full sponsorship. It was difficult, with a superstar, to gain agreement to do the entire commercial package rather than just lead-ins. At that time, in 1960, a lot of Hollywood's most scintillating personalities still viewed TV as a crass interloper that would, perhaps, go away.

Those planning the big event knew that there was no way Kaye and his former wife, writer Sylvia Fine Kaye would even consider commercials unless they had a clear idea of exactly what was planned. The agency wrote this note to Mrs. Kaye:

> Because we must present nine cars in a single commercial, that commercial has to have entertainment value par excellence. People's reactions to the show will be tempered by how they feel about the advertising.

The complications of writing, storyboarding, and getting approval from the corporation, each of its five divisions, Danny Kaye's agent, Kaye himself, and his wife consumed the months from February through August. With the entire hour having to be considered, commercials weren't uppermost in the minds of the star, the writer, or the show director, Norman Jewison. At the advertising agency in Detroit, and in New York and Hollywood, half a dozen people were involved in writing and rewriting the commercials and in the myriad details of production. Color TV was relatively new, and cars from the various makers (Cadillac, Buick, Oldsmobile, Pontiac, and Chevrolet) had to be color-coordinated. Studios had to be reserved in both Brooklyn and Hollywood pending agreements as to where the show would be videotaped.

In May, some 100 days after the project was launched, McHugh, Gail Smith, and others from the agency and client attended a vital meeting in the offices of Lew Wasserman, head of MCA. General approval of the agency's creative approach was voiced, time periods for the show were agreed upon, and the relative cost of color versus black and white was discussed.

The next day, commercial scripts were reviewed with Danny and Sylvia Kaye, and things went smoothly. Sylvia wanted time to consider the commercials and their effect on the whole show. There was still some question as to whether use of the star was wise. That question, along with a number of others, was finally and irrevocably settled in August. Production took place in October, and the show aired on Sunday night, October 30, 1960. Was all the effort related to sole sponsorship and star integration worthwhile?

Here are excerpts from *Advertising Age* columnist Harry McMahan's follow-up story:

> Raves on the Danny Kaye show have just about topped anything the program critics have seen since Fred Astaire's similar debut on television. . . .
>
> But this time the reviewers raved about the commercials as much as the program. It was a two-way smash for a one-man show.
>
> Commercials were spotted well into the hour and there were only two. The first an astonishing 4½ minutes that never let up, the second a rapid-fire Danny Kaye patter song that clipped it off in the remaining 1½ minutes.
>
> Kensinger Jones, C-E's senior vp and creative director, authored the big one personally, and only a stopwatch could tell you it ran as long as it did. There were nine cars featured, yet the whole thing was as simple as opening a door. And as smooth as Hydramatic.
>
> It opened with Danny telling a story to children: "Once upon a time there were five cars . . ." and before he got through he'd covered all nine, carefully positioning each in the GM spectrum. For instance, Danny explained the two Buick lines and the new Buick Special this way:
>
> "It's for people who want all things that a Buick is . . . but in a smaller package . . ."
>
> Simple. Wonderfully human language. A superb commercial.

The other commercial on the show packed more into it than (or at least as much as) the fine Federal Express commercial hailed as one of 1983's best. Kaye, who is and was famous for his rapid-fire delivery, machine-gunned lyrics like these:

> If you want to be seen in a new limousine, or if you're a fan of a new sedan, if you fancy a wagon, a new station wagon, you feel you want steel or aluminum . . . a four-door, a two-door, a black door, a blue door, a small car, a tall car, GM has a lot and they all of them got plenty of room-in-'em.

It was a joint effort by Campbell-Ewald's Fred Lounsberry and Sylvia Fine Kaye, and it was as much of a show-stopper as any of the program numbers.

DRAMATIC RESULTS

Research, as well as trade reports, attested to the extraordinary effectiveness of show sponsorship and integrated commercials. Prof. John Dollard of Yale had developed a research instrument that measured "National Learning Scores," which, he said, "is the best estimate we can make of the total good rendered the sponsor by the commercials and the carrying show. General Motors reaped a National Learning Score on the combined commercials far beyond any we have previously recorded. The Kaye commercials did three times as well as the average."

Social Research, Inc., also measured the results of the total effort. Some interesting findings were reported: "Viewer interest and involvement was as great during the commercials as during the performance itself."

Direct quotes from respondents bore this out:

> I really liked it. He made you feel the cars were really the best cars made.
>
> I liked the way they showed the old cars and Danny told the kids about them. Then he showed the new GM cars and the compacts.

An interesting postscript to this story is the December 1983 report that Campbell-Ewald was considering placement of a series of stage dramas in prime time on the USA Cable Network for Chevrolet—an $850,000 cable buy.

THE FUTURE OF SPONSORSHIP AND INTEGRATED COMMERCIALS

Is it possible that the sixties were the brightest and best years of network television; that it reached heights as a purveyor of great entertainment and an effective commercial vehicle that it will never scale again?

Could be.

Time magazine advertises "The $30,000 can of beer" (Figure 4-3) and says, in a subhead, "Almost 40% of the average prime-time television audience leaves the room during commercial breaks. If you sell cars, you simply can't afford to talk to the walls."

In copy, it drives this nail home: "At $84,000 for a typical prime-time 30-second spot, you're spending $32,000 advertising to empty space."

The fact that people are leaving the room can be construed as a vote against network programming as well as commercial content. Cable has the opportunity, through selective programming and integrated commercials, to keep them in their seats.

Some advertising agencies are convinced enough to have revived the programming function. Compton Advertising is packaging "The Great American Homemaker" on the USA network for Procter & Gamble. D'Arcy-McManus & Masius Inc. is involved in special sports programming for its client, Budweiser. There is increasing evidence that agencies have an important part to play in recapturing the selling power of proper sponsorship for their clients.

William Wiener of BBDO says his agency is getting away from spots. Clients want something they can call their own, program association or program "ownership."

CONTROLLED COMMUNICATION

One fascinating experiment made possible by cable's relatively low cost is the American Express in-house-produced series, "How's Business?" It aims directly at the businessperson.

Management must view it as an institutional or public service venture. In tone, it can be said to advocate a better understanding and appreciation of business institutions. For instance, the first two

Figure 4-3 The erosion of network television audiences is dramatized in this ad in *Time* magazine. Cable is used by some advertisers to fill the viewership gap. (Courtesy: Time, Inc.)

shows dealt with employment of the handicapped and business support for the arts.

The Satellite Program Network of Tulsa cleared time, and Amex had a vehicle and a channel to reach (1) the public, (2) employees, (3) stockholders, and (4) vendors, who were especially alerted to the time and channel for viewing the show.

With 388 cable affiliates and a potential audience of 5.4 million, SPN seems to offer a reasonable chance at most of this targeted audience. Given proper promotional support, the show can also reach a useful and valuable cross-section of academics, business leaders, and government people.

All this for costs estimated to be under $50,000 would seem to indicate new possibilities, through cable, for the presentation of ideas as well as merchandise and service.

The importance of the narrowcasting concept cannot be overemphasized when cable is being considered as a means of controlled communication with a selected audience. Combined with direct mail or print advertising in cable guides, it can be a "by invitation" chance

Figure 4-4 Program underwriting logo. Audio: "PROFILES OF NATURE is made possible in part by the Board of Water and Light, Lansing's customer-owned utility, serving our community for 100 years with water, steam and electric power." (Board of Water and Light, Lansing, Michigan)

to hear experts speak on any number of issues affecting business. Some corporations have bought specific time segments on SPN to present a visual "annual report." Although such programming isn't of broad general interest, it certainly is meaningful to a company's stockholders, employees, customers, suppliers, and even those who question the company's operations. If cable can provide a platform for an interesting, graphic, and meaningful presentation of an institutional message and can do so at an affordable price, it is providing a tool that did not exist previously. The unanswered question is how many organizations—national, regional, or local—will have the ingenuity and courage to use it.

UNDERWRITING

It also seems possible that business can participate financially in making better use of cable access and local origination channels, just as it does in underwriting programs on both radio and television for PBS (Public Broadcasting). Grants from companies like Hewlett-Packard and Mobil have been welcomed, and all the companies have expected in return is acknowledgement of their assistance. The value is in terms of goodwill.

Isn't that same kind of value achievable by a local bank, a regional utility, a chain of retail establishments, if they underwrote a series of cultural and educational programs on appropriate cable systems in their particular service areas, or funded the telecasting of city council or school board meetings?

It will take some thinking, some discussion, some investment to make it happen. It will require support from the cable system, the advertiser, and the agency, and acceptance of the idea by access channel coordinators.

Chapter 5

Buying and Scheduling Cable Advertising

Like any advertising medium, cable has particular characteristics that dictate its use. On the surface, cable appears to be most like broadcast television, because it involves the transmission of video images. In terms of the audience, however, cable television may be more like radio, because its audiences tend to be relatively small and specialized.

Cable television advertising is also similar to newspaper advertising, in that it can provide strong retail and merchandising identification. Furthermore, cable, like newspaper space, is often sold on the basis of the number of subscribers available. Cable television advertising can also be described in terms of traffic advertising, such as outdoor and transit advertising, when automated channels and program guide channels are considered. Instead of traveling through the community for exposure to the advertising messages, the viewer is moving through the cable channels. The advertising messages need to be short, and they must communicate very quickly, much like the outdoor media.

The analogy to radio is especially strong in terms of buying and scheduling cable advertising. Don Tennant, one of the three original members of the Leo Burnett Television Creative Department, likens cable TV to FM radio and thinks that its development parallels that of the all-sound medium. FM radio was a highly selective vehicle in its early days, with relatively few commercial interruptions. Few radio sets were equipped to receive FM, just as a large number of TV sets today have not been wired for cable. As FM continued to attract a growing audience, and as FM receivers became commonplace, FM stations began to carry more advertising. Programming became very specialized, particularly in the big markets. There were stations carrying classical, country, rock, hard rock, bluegrass, big band, and other varieties of music. FM radio became an invaluable medium for local advertisers in terms of efficiency, because the range of each station was limited and costs were comparatively low.

Today, FM is part of almost any radio buy, and there are thousands of advertising-supported stations. Stations were formerly bought for their special-interest aspects, whereas they are now more and more selected on the basis of the audiences that they can deliver. Cable seems to be following the same path, but with pictures. Hence the media strategies that should work best in cable will presumably be similar to the strategies that work best in radio.

Figure 5-1 Don Tennant, who heads his own agency in Chicago, sees cable advertising parallelling FM radio in its development. (Courtesy: Don Tennant Co., Inc.)

WHAT CABLE CAN SELL

Any strategy using cable television advertising depends upon the nature of the time available from the local cable system. As discussed before, the cable operator may offer time on advertiser-supported satellite networks or on local origination programs. Advertising might also be found on audio and text channels. The satellite networks, such as ESPN, MTV, CNN, and USA, are the dominant source of available local advertising time in cable. Although in some markets the local origination channels have contributed to advertising success, in most markets, advertising on local origination channels is considerably less developed than on the satellite networks.

What the operator can sell is termed a "spot." The spot is the time that the commercial message actually takes. In broadcast television, the overwhelming majority of spots are 30 seconds long. Spots in cable, as in broadcast television, can be a variety of lengths, including 10, 20, 30, 60, and 120 seconds. Longer spots are also possible. Named "infomercials," these are discussed elsewhere in this book.

The spots a cable operator has to sell are referred to collectively as the spot inventory. A typical satellite network gives the local operator two 30-second spots per hour of the twelve openings that are created for commercials. If the network programs 24 hours, then the cable operator would have approximately 48 spots per network carried daily. The average cable system selling spots on six cable satellite networks would have an inventory of nearly 300 satellite spots daily and another 300 on a local orgination channel.

Any local origination channel would add considerably to the available inventory because its spots would not have to be shared between the network and the local operator. All spots would belong to the operator. However, those systems that put on successful local origination channels have made a considerable investment in the programming and support staff. Many cable systems have thus far been unwilling to make the investment. They argue that all the broadcast and satellite channels carried on the cable attract almost all the viewers. Competing would be tough. The attitude seems to be to sell what one has first, before creating anything else.

WHO BUYS LOCAL CABLE

Most cable operators report that the overwhelming majority, usually over 90 percent, of the advertisers who buy the local availabilities are local advertisers. A local advertiser is defined as an advertiser who has a retail establishment in the local service area. (see Figure 5-2) The remainder of the local inventory is sold to regional advertisers, except for a very small portion—usually 2 percent or less—if any at all, sold to national advertisers. An exception is Manhattan Cable, a very large system with high visibility among the New York advertising agencies.

A substantial portion, or between one-half and three-fourths, of the spot inventory for most cable operators goes unsold. As would be expected, cable network prime time local availabilities and major cable network sports events may sell out completely, and the spots running very late at night or early in the morning have the highest portion unsold. Cable is much like broadcast television and radio in that the spots during the peak audience times are the easiest to sell.

Figure 5-2 Cable reaches prospects for local advertisers. (Coaxial Communications, Columbus, Ohio)

Advertisers may have to contract over a long term with the cable operator to reserve time in the most desirable programming. Despite the reluctance from many cable operators to develop much local origination, most systems that run local sports events on their local origination channels report the sports events often being completely sold out at the most desirable times.

RATE CARDS

A *rate card* will show the asking price of individual spots, as well as any discounts that may be offered. Most cable operators have only one rate card—that is, a single base rate offered to all advertisers. As the medium grows, some operators may go to a *dual-rate structure*, like that of broadcast television and radio, that offers a higher rate for national advertisers and a lower rate for retail or local advertisers. This is due in part to the presumption that a national advertiser buying the local spot from a sales representation firm on a local cable system would purchase through an advertising agency. Both the representation firm and the advertising agency would then take a commission. Local advertisers would most likely be placing their advertising directly with the cable operator.

Rate cards vary substantially in their complexity from one cable system to another. Any two cable systems are likely to have different rate cards, since the structure of the rate card tends to be peculiar to the individual operator. However, certain common elements are almost always present on all rate cards. Generally, the base rate is tied to an estimate of the audience delivered. In other words, high audience times should have a higher base rate than lower estimated audience times. A problem for cable operators has been arriving at reasonable estimates of the audience. Because of this, some operators make no distinction between high and low audience times but offer a single base rate for all times.

Another way an operator might determine the base rate for a spot would be consideration of the demographic characteristics of the audience delivered. Generally speaking, audiences of men are difficult to deliver and consequently cost more than audiences of men and women combined or women alone. An operator might, for example, have a higher base rate for a spot on ESPN (the Sports Network) delivering a predominantly male audience than for a spot on CNN delivering an audience of both men and women.

Given a base rate for a particular audience, there are generally two forms of discounts. The first is a discount for *volume* or *frequency*. This discount applies to the number of spots purchased during some specified time interval, such as a week. This means that the rate per spot would be lower as the number of spots purchased per week increases. Generally, some kind of step scale would be shown on the rate card. The second type of discount is a discount for *continuity*. This discount lowers the rate paid per spot based on the number of weeks the schedule is run. Buying spots each week for some minimum number of weeks would result in a lower per-spot cost. Discounts for continuity help ensure the cable operator of a dependable source of advertising revenue. Many cable operators would offer discounts for both volume and continuity. Like other businesses, operators may also offer discounts for cash or early payment.

Rate Card Example

Figure 5-3 is a typical rate card from a cable operator. All the elements are shown. Continuity discounts would be shown over number of weeks, usually in multiples of 13 weeks. Discounts are shown at the bottom of the first part of the rate card. For 13 weeks, the advertiser gets a 10 percent discount, for 26 weeks 15 percent, and for a year or longer, 20 percent. The second part of the card, labeled "Bulk Plans," shows a typical volume discount structure. As the number of spots increases, the rate per spot decreases. In the first line, the 75-time rate is $28 per spot, which compares to the 600-time rate of $22 per spot. The rows, 1, 2, and 3, represent different seasons. The 1 is the holiday season, hence the highest viewing, and the 3 is the summer months. When an advertiser purchases time on a grid, the particular program or even channel purchased may be unknown to the advertiser; only the audience that should be delivered is known.

A grid makes "rotations" very simple for the cable operator. A rotation means that the spot will be moved from the same time position to another time position with an equivalent audience. There are two kinds of rotations, horizontal and vertical. Vertical rotations generally refer to movements at different times during the same day, whereas horizontal rotations refer to the same time across different days. What is important in a grid system is that the same level of audience is maintained.

This rate card introduces two other elements often found in broadcast rate cards, the "grid" and the "plan." Advertising time can be offered in one of two ways, by time of day or by size of audience. The traditional means of offering time in broadcast television has been by time of day; that is, the station offers the spot at a fixed time, regardless of the audience. The modern method is the grid, which recognizes that advertisers are purchasing audiences and not times. A grid value is determined according to an estimate of the audience, so that the advertiser is buying an audience. In Figure 5-4, the AAA, AA, A, and B represent different audience estimates, with the AAA bing the largest.

Commonly offered to an advertiser is a plan or package plan, which combines the high audience times with some of the low audience and hard-to-sell times. A plan is a common way for a cable operator to sell spots that might otherwise be difficult to sell. The second part of the rate card illustrates a bulk plan.

An additional concept, "preemptible," is also introduced by this example. Preemptible means that if the cable operator can sell the time at a higher rate, the time will go to the higher-paying advertiser. An advertiser may be willing to take a chance on preemptible spots to get a bargain in price. The ROS, or run of schedule, will be discussed in the next section under placement.

The rate card is merely the cable operator's asking price for the spots. Many cable operators negotiate the rate. The degree to which negotiation can occur, of course, will depend upon the particular cable operator and the number of unsold availabilities.

The ways of negotiating are numerous. Some operators attempt

GRID RATE CARD

Call Continental Cablevision for prevailing grid

Rates:

Grid	AAA	AA	A	B
1	$60	$45	$30	$15
2	$55	$40	$25	$10
3	$45	$35	$20	$ 5

Term discounts for above rates only.
Term: 13 weeks — 10%, 26 weeks — 15%, 52 weeks — 20%.

Bulk plans: R.O.S. Preemptible Minimum Buy

	75x	150x	300x	600x
1/3 AAA, 1/3 AA, 1/3 A	$28.00	$26.00	$24.00	$22.00
1/6 AAA, 1/3 AA, 1/3 A, 1/6 B	$23.50	$21.50	$19.50	$17.50
1/2 AA, 1/2 A	$22.40	$20.40	$18.40	$16.40
1/3 AA, 1/3 A, 1/3 B	$18.40	$16.40	$14.40	$12.40

All above :30 second rates
R.O.S. = run of system

Continental Cablevision 485-6547

Figure 5-3 Rate card (Continental Cablevision of Lansing)

PROGRAM CLASSIFICATIONS

Network	Program	Class
ESPN	Football, Basketball & Major Sports Playoffs	AAA
	Major Sports: Baseball, Hockey, Boxing, etc.	AA
	Talkshows — Sports	A
	Sportscenter — Sportscenter Plus	A
	Minor Sports	B
USA NETWORK	Football, Basketball & Major Sports Playoffs	AAA
	Major Sports: Baseball, Hockey, Boxing, etc.	AA
	USA Daytime (non sports)	A
	Talkshows (sports)	A
	Minor Sports	B
	Night Flight	A
MTV	**Monday — Saturday**	
	12:00-3:00 p.m.	A
	3:00-9:00 p.m.	AA
	9:00-12:00 a.m.	AAA
	Sunday	
	12:00-6:00 p.m.	A
	6:00-12:00 a.m.	AA
	Saturday — Sunday	
	Movies and Concerts	AAA
CNN	12:00 6:00 p.m.	B
	6:00-7:00 p.m.	AA
	7:00-10:00 p.m.	AAA
	10:00-12:00 a.m.	AA
NASHVILLE	12:00-6:00 p.m.	B
	6:00-12:00 a.m.	A
LIFETIME	12:00-6:00 p.m.	B
	6:00-12:00 a.m.	A

Local Weather Channel/Shoppers Guide (Videotext)

Eight-line message with NOAA weather line & time/date line above, local weather line below, displayed 24 hours a day.

Weekly Rate	$15/day
Monthly Rate	$12/day
Yearly Rate	$10/day

Figure 5-4 Program classifications (Continental Cablevision of Lansing)

to hold to published rates, but they actually "bonus" advertisers. The bonus is the promise of additional spots from unsold inventory. The extent of the bonus spots can be formally stated. The amount of discount for volume or continuity is also negotiable. Furthermore, many cable operators are willing to put together a special package of spots tailored to the needs of an advertiser. This package may not conform exactly to the plans and discount structures presented in the rate card, so the price must be negotiated. Merchandising (cable guide advertising, point-of-purchase materials, billing stuffers) may also be part of the negotiation. Certainly negotiation should be an important part of the transaction with the cable operator.

PLACEMENT

To take advantage of programming possibilities on cable, media planners need to thoroughly analyze what is available. Then media buyers will have to be completely specific.

Most cable operators, just like operators of the other media, will offer two methods of placing spots. The first, allowing the lowest price per spot, would be *run of schedule*, or ROS. This allows the operator to run the spots during high audience times, or low audience times, or on any of the available programs or services. Often the arrangement guarantees proportional representation of spots in several classifications of time based on expected audience levels. This is in contrast with a fixed placement, where the advertiser pays a premium rate but is guaranteed placement within a particular time or a particular program. Some operators may also offer a compromise, by which the flexibility given the operator in moving the spots through the schedule is within some defined limit, such as sports programming. In general, it is in the interest of the cable operator to promote ROS sales, because they move the less-desirable spots that are in low audience times.

Buying ROS gives the cable operator the control over when the spots will run. This is wise for the cable advertiser only if the ROS rates are substantially lower than the fixed-placement rates. ROS advertising, because it serves the convenience of the seller, should be sold at less cost per thousand viewing households than the cost per thousand viewing households in a fixed-placement purchase. If audience data are available so that this comparison can be made, the advertiser may get more for his or her money in ROS. But the audience numbers are seldom available, so most local cable advertising time is purchased on the basis of logic. This being the case, it is much easier and safer to select particular programs on a fixed-placement schedule.

COSTS

Cost is also a critically important factor in developing any advertising media strategy. (See Figures 5-5 and 5-6.) To compare differing media vehicles, cost is calculated as cost per thousand, or CPM. In

Page Display Advertisement

The Weather Channel

THE WEATHER CHANNEL

The crawl message can include up to 198 characters. Local forecasts will appear one minute out of every five. Your crawl message will appear at the bottom of the second, third, and fourth pages of each local forecast for a total of 36 seconds. The message will repeat twelve (12) times per hour, a total of over 7 minutes each hour.

Warren $250.00 per week (exclusive) $750.00 per month
Clinton $125.00 per week (exclusive) $375.00 per month

Cable Advertising For Profits.

Crawl Messages on Billboard Channels or Full Page Displays

Crawl Mesages — 240 characters
Full Page Displays — 8 lines/32 characters each

Warren	Clinton	Sterling Heights
Weekly - 75.00	Weekly - 50.00	Weekly - 67.50
Monthly - 200.00	Monthly - 135.00	Monthly - 175.00

Cablemate Guide

	1 Time	6 Times	12 Times
Inside Full Page	825.00	800.00	775.00
Inside 1/2 Page	485.00	460.00	440.00
Inside 1/4 Page	250.00	245.00	240.00
Inside 1/8 Page	135.00	130.00	120.00

Rates subject to change.

COMCAST CABLEVISION
S.E. Michigan Regional
6095 Wall Street, Sterling Heights, MI 48077
978-7739

Figure 5-5 Rate card (Comcast Cablevision)

COMMERCIAL RATES

Warren All ROS

1-14	- 25.00*	50-74	- 17.50*
15-24	- 22.50*	75 plus	- 15.00*
25-49	- 20.00*	*30-second spot	

Clinton All ROS

1-14	- 15.00*	50-74	- 11.00*
15-24	- 13.00*	75 plus	- 10.00*
25-49	- 12.00*	*30-second spot	

Sterling Heights All ROS

1-14	- 22.50*	50-74	- 15.00*
15-24	- 20.00*	75 plus	- 12.50*
25-49	- 17.50*	*30-second spot	

- Any two (2) system combination buy — use rates less 10% discount.
- The three (3) system combination buy — use rates less 15% discount.
- Any special insertion accommodation other than ROS will require special rate consideration.

Figure 5-6 Rates for text messages and the printed program guide (Comcast Cablevision)

general, cost per thousand is determined by taking the cost of the spot, multiplying by 1,000, and dividing by the size of the audience delivered. The problem for cable advertising has been in making a good estimate of the size of the audience. An advertiser should calculate approximate cost per thousand for all the media vehicles that might be considered. For a local advertiser, this would mean comparing the cost of the local radio station, the local television station, the newspapers, local outdoor and transit media operators, and, insofar as possible, the local cable operator. The following example shows how this cost comparison can easily be made using a hypothetical community.

CALCULATING MEDIA COSTS

For newspapers:

$$\text{CPM} = \frac{\text{Cost of ad} \times 1{,}000}{\text{Circulation}}$$

For broadcasters and cable:

$$\text{CPM} = \frac{\text{Cost of 1 unit of time} \times 1{,}000}{\text{Number of homes reached by a given program or time period}}$$

Computations for Anywhere, U.S.A.:

Anywhere Gazette, ¼ page in the morning edition:

$$\text{CPM} = \frac{\$660 \times 1{,}000}{115{,}000} = \$5.74$$

Anywhere Radio WXXX, 30-second spot during morning drive time:

$$\text{CPM} = \frac{\$70 \times 1{,}000}{22{,}000} = \$3.18$$

Anywhere Cable Television, 30-second spot during "AAA" time:

$$\text{CPM} = \frac{\$50 \times 1{,}000}{15{,}000} = \$3.33$$

QUALITATIVE FACTORS

Cost per thousand, however, cannot be the primary factor considered in cable advertising until authoritative audience numbers are available. Qualitative factors must be considered. Some advertising agencies continue to rely on traditional methods of media purchase for this nontraditional medium and, consequently, do not recommend it. Meanwhile, their clients see their competition using cable effectively and therefore may pressure agencies into considering cable on an experimental basis.

An example is Delbert M. Lothes, president of Buckeye Optical Company, an advertiser on the Coaxial Cable System in Columbus, Ohio. He has an in-house advertising agency, so media decisions are his own. Using television, radio, and print, he spends 7 to 8 percent of his net sales on advertising. Currently, he allots 5 percent of his total TV budget to cable. Ordinarily, the same commercials are telecast on regular and cable TV and are purchased from a packager of syndicated materials for use by optical houses. However, one locally produced cable commercial has given him something to think about. The quality and low cost of the effort has Lothes considering more local production. For example, because research indicates that two-thirds of the purchases of eyeglasses involve women, he can pick appropriate programs from the cable schedule that should attract women viewers and create special commercials for them.

Another advertiser, BancOne, began advertising on the Warner-Amex cable system in Columbus in an effort to reach prime prospects for its banking services in upper-income suburbs. Having no statistics, the firm set out to reach its target audience with more frequency and for less cost. To explain things like automatic banking, a longer infomercial with complete information on how it worked was used. The effectiveness of the program led BancOne into additional advertising on other cable systems.

CABLE AUDIENCES

The size of the audience for a cable program is described in the same way as is the audience for broadcast television and radio. The fundamental audience measurement for an advertiser is a *rating*. A rating is a proportion of the viewers of a particular program at a particular time compared to the total potential audience. In the case of television, the total potential audience would be all homes or persons with working television sets. In the case of radio, it would be all persons with working radios. As a practical matter, for both television and radio, virtually everyone in a given market area has either a television set or a radio, so that the rating is really nothing more than a percentage of the total market. A broadcast television rating of 20 percent then means that one out of five homes in the market was viewing that program. The convenient thing about ratings is that they are additive, because the base always remains the same—that is, the total market.

Because it is often easiest to calculate ratings on the basis of households, ratings are usually expressed in terms of households. This is especially true when the rating is for the overall market without regard to any demographic characteristics. Obviously, households do not watch television, people do. People have demographic characteristics, such as age and gender. Ratings can also be calculated on the basis of people rather than households. Household numbers should not be combined with people numbers in the same calculation, however, since the average household contains more than two people.

Generally, broadcast television is sold in terms of ratings, or aggregated ratings, known as *gross rating points*. If an advertiser purchases 100 gross rating points per week, that means that multiple spots were purchased with ratings that total 100. This could be, for example, ten spots in shows with ratings of 10 percent, or five spots in shows with ratings of 20 percent.

The use of ratings to measure cable audiences presents some very special problems. Although television and radio enjoy virtually 100 percent penetration into most markets, cable does not. Typical cable television penetration is in the neighborhood of 50 percent. A rating of 10 percent for a cable television program in a typical market would mean that one out of ten homes in the market viewed the program, but the same rating based on cable homes only would mean one out of five cable subscribers, assuming a 50 percent cable penetration. To be consistent, ratings should be used by advertisers in the same way for cable television as they would be used for other advertising media. The fact that cable may reach only half the homes in a given market does not necessarily create any special problems.

Local newspapers, for example, generally do not enjoy more than 70 percent penetration by homes in a given area, and even a lesser penetration of the total population, because not everyone in a home reads the newspaper.

In cable, it is especially important to know the base for a rating. The base may be all television households, whether or not they are cable subscribers. Or the base may be only the cable homes. Another base may be individuals—for example, "males 18+ in TV households," or "males 18+ in cable households."

The typical rating for a cable program would be relatively low compared to a typical rating in broadcast television. The ratings for cable programs would in general be most comparable to the ratings found in radio. This means that for a schedule similar to one that would be run on broadcast television, a large number of spots should be scheduled during the week, similar to a typical radio schedule. A typical weekly television schedule, such as 100 gross rating points, might require as many as 40 or more spots per week on cable programs, whereas it might require as few as five on broadcast television.

The Cable Television Advertising Bureau reports that all the advertiser-supported satellite networks, including the superstations,

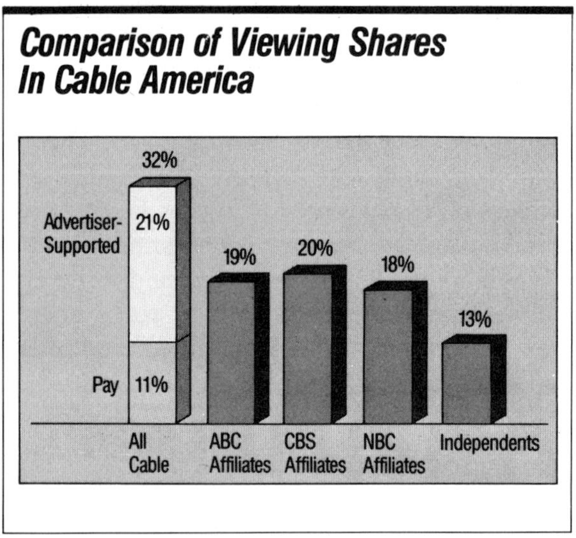

Figure 5-7 In cable homes viewing of all advertiser-supported cable networks exceeds viewership of any single broadcast TV network. (Cabletelevision Advertising Bureau)

together enjoy an 18 percent share of subscribers. With cable audiences growing, it could pay local advertisers and agencies handling their accounts to have a look at what is available, in their area, on the cable networks that are advertising-supported.

MEDIA STRATEGIES

Cable, like all the other advertising media, must be employed strategically in order to maximize the benefit to the advertiser. The most fundamental media strategies and objectives—reach, frequency, and continuity patterns—apply to cable just as they do to other media.

Reach

Reach is perhaps the most fundamental media objective. *Reach* is defined as the target prospects who are exposed at least once to an advertising schedule. It is the unduplicated audience and can be expressed either as the number of prospects or as a proportion of the total universe of prospects. In most cases, it is extremely difficult to achieve a reach of more than 90 percent regardless of the size of the advertising budget or the media used.

The concept of reach is important in comparing one medium to another and judging how many different people are exposed to the advertising message. Cable may reach some people who are not easily accessible through other media. For instance, the viewer of "Business Times" on cable network ESPN may watch very little other television. An 18-year-old college student may not have time for the print media and broadcast television, but may watch MTV in spare moments while doing other things. So reach should be considered in terms of real prospects who can be segmented through the narrowcasting of cable.

To build reach in cable, it is advisable to purchase as many different kinds of programs as possible, not just a single category, such as sports. Greater reach would require considerable diversity in the programs purchased. For example, ESPN would be purchased along with the other networks.

Effective reach is another important media concept. Effective reach refers to the number of exposures necessary to achieve the objective for the campaign. Effective reach is usually believed to be between three and ten exposures through the period during which a purchase is under consideration by the consumer. Using this as a guideline, the appropriate spending level can be easily determined.

No doubt several factors influence the effective reach, not least of which is the level of competitive spending. As more competitive spots are seen by the prospect, more exposures will be needed to offset them. Other factors would include such things as the complexity of the message and the novelty of the product or service. Advertisers must make some judgments to determine where in the frequency spectrum they should fall.

Frequency

Frequency is defined as the average number of times a target prospect who is reached is exposed to a message. Frequency, or average frequency, is calculated by dividing the total gross audience by the reach. The total gross audience is normally expressed as the total gross rating points, or GRPs, for the schedule.

As more spots are purchased, increasing the number of gross rating points, either the reach will be increased or the frequency will be increased, or both. Reach increases at a decreasing rate, so that as the reach approaches the maximum, additional spots will be more likely to produce additional frequency rather than additional reach.

Most of the ratings for cable programs tend to be relatively low, especially when compared to prime time network television. In general, when a schedule consists of many low-rated programs, it will tend to build frequency rather than reach. This makes cable similar to radio in that it is easier to build heavy frequency than to achieve extensive reach.

Continuity

Continuity refers to the consistency of advertising placement through time. There are essentially three strategies: a continuous strategy, when the advertiser maintains the same level all the time; a *flighting* strategy, when the advertiser will advertise for a short period of time and then not at all; and a *pulsing* strategy, when the advertiser periodically increases the effort but is always advertising at some minimal level.

The two rationales for either a pulsing or a flighting strategy are either to stretch a limited budget or to avoid overadvertising, or "wearout." Placing too many exposures against the target audience will certainly lead to diminished effectiveness, called advertising wearout. Whether or not the wearout phenomenon actually exists is controversial. Some studies clearly show the wearout effect, others do not. Most studies, however, do demonstrate lessened effectiveness as frequency levels increase.

Flighting and pulsing strategies, when systematically studied, prove to be best for second-place advertisers, generally those with approximately a 20 percent share of the market. Dominant advertisers, those with a 40 percent share or more, are better off if they maintain a continuous schedule. Smaller advertisers, those with a 10 percent share or less, have not been systematically studied, but they may have to adopt a flighting or pulsing strategy because of limited budgets, regardless of other factors.

COMPLEMENT TO OTHER MEDIA

Cable can be used as a primary medium for the retail advertiser, or it can be used to supplement or complement other local advertising media. Cable has many of the audience characteristics of radio, but it still has the effect of television. Part of the interest in cable has been its ability to attract audiences away from the traditional broadcast television networks. Certainly broadcast audiences have declined since cable penetration, and cable programming has increased as shown in Figure 5-9.

An advantage, then, of using cable advertising is as a defense against the erosion of the broadcast television audience. By using cable to supplement a television schedule, an advertiser can be assured of maintaining the desired audience.

Cable can also be used in coordination with other local media. A newspaper promotion could be especially useful in drawing attention to special cable programming and infomercials. Advertising on cable, on the other hand, could also be used to draw attention to coupons or sales appearing in the newspaper. Similarly, advertising on broadcast television and radio can be used to refer to special advertising and other programming appearing on cable.

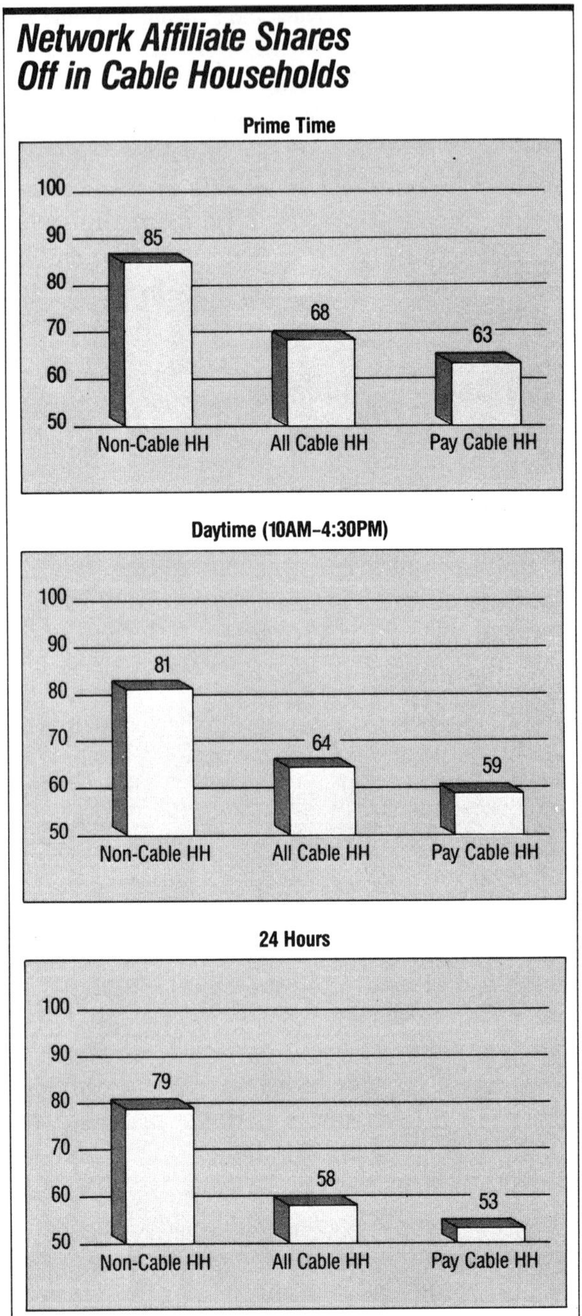

Figure 5-8 Cable's share of available viewers is significant. (Cabletelevision Advertising Bureau)

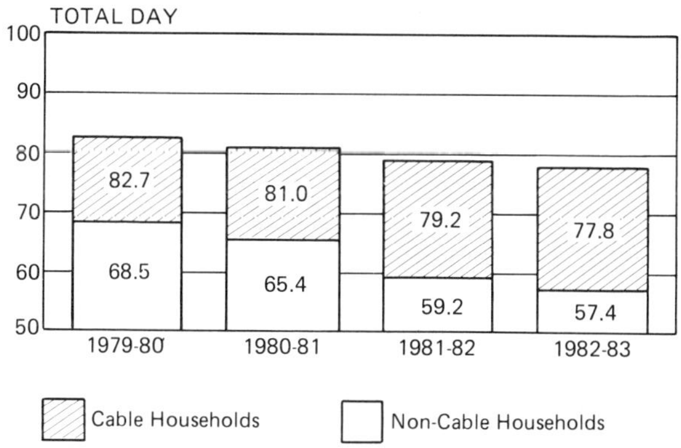

Figure 5-9 Decline in cable viewership is less than half that of traditional TV. (Cabletelevision Advertising Bureau)

ROADBLOCKING

A fear many advertisers have in television, and especially cable, is the phenomenon known as "zapping." Zapping refers to the changing of channels when the advertising appears, thereby reducing the size of the audience for the advertising.

A strategy to counter this, *roadblocking*, can be used in cable. The concept of roadblocking involves buying the same spot on all available channels during the same time period. This means that the viewer changing channels will see the same spot. This strategy also has the effect of accumulating audience, so that the rating for each show will be added to the total audience delivered. Cable channels must be included in the advertiser's schedule along with the broadcast channels for this strategy to work.

INTERCONNECTS

A particularly difficult problem for cable operators in selling spots to advertisers is the geographic audience they have available. Many larger markets, for example, are divided into areas with different operators serving them. Suburban communities may be served by different operators from those in the central city. If an advertiser defines a particular market area in traditional terms, it may be necessary to contract with many cable operators to cover the market. This can be both confusing and difficult for the advertiser.

Buying a market like this can be simplified through interconnection. Essentially, there are at least three reasons for operating an interconnect: to facilitate advertising sales, to spread the cost of expensive local programming, and to provide business and institutional communications. The interconnect can cover a geographic market area, a state, or some other geographic territory. The specific territory covered by an interconnect is of interest to the advertiser.

There are two types of interconnects, the "soft" interconnect and the "hard" interconnect. The soft interconnect is an interconnect on paper, with tapes being delivered to each cable system. The hard interconnect has a physical connection, usually microwave. Many of the soft interconnects are planning hard interconnects in the future. Approximately half the interconnects in the country are currently soft.

The Cable Television Advertising Bureau has 38 interconnects listed in its directory. These interconnects represent approximately 24 percent of the cable subscribers and just over 7 percent of the cable systems. The largest interconnect is in New Jersey, as shown in the table below which lists the top 20 interconnects by the number of subscribers.

There are several examples of advertising interconnects that cover a fairly large geographic area. One of these is the Centel Videopath Network which covers most of the Chicago area. The network consists of microwave links between the headends of 12 separate cable systems. Most of the cable operators have multiple systems in more than one municipality. The network itself is shown in Figure 5-10 which includes all of the headends and two switching facilities located in Des Plaines and Wheaton. The central hub site for the microwave network is located at McCook. Other interconnects can be found in New Jersey, Philadelphia, New York, San Diego, San Francisco, Seattle, Boston, Cleveland, Connecticut, and many other markets.

Many problems prevent widespread availability of advertising oriented interconnects. Rarely would adjacent cable systems be equivalent. For example, one system may be a 35-channel system carrying five programming networks; another system may be a 12-channel system carrying only one; and yet another may be a 108-

TOP 20 ADVERTISING CABLE INTERCONNECTS

		(000)
1	Cable Television Network of New Jersey	976
2	New York Cable Advertising Network	885
3	Bay Area Cable Interconnect	573
4	Boston Interconnect	540
5	Cleveland Interconnect	374
6	Cable Adnet, East Pennsylvania	334
7	San Diego Interconnect	325
8	Connecticut Cable Interconnect	312
9	Northwest Cable Interconnect	270
10	CBS Interconnects, Chicago	260
11	Detroit Interconnect	245
12	Suburban Buffalo Cable System	220
13	Southeast Florida Interconnect	194
14	Houston Interconnect	185
15	Atlanta Cable Advertising Network	172
16	Mountain Cable Network	164
17	New Orleans Interconnect	163
18	Tidewater Cable Interconnect, Norfolk	163
19	North Carolina Cable Advertising Network	161
20	Greater Dayton Cable Interconnect	149

Cabletelevision Advertising Bureau

channel system carrying all the available services. Administrative policies, available headend equipment, and other problems make connection sometimes difficult. As a practical matter, interconnects remain an important future consideration for the cable advertiser. There is little doubt but that they will become a widespread and viable mechanism in the near future.

One solution to the interconnect problem is the one used by the Centel Videopath Network. Operating as a common carrier, Centel contracts with CBS Interconnects which is the advertising sales agent. Incidentally this interconnect is listed as number 10 in the Top 20 CBS Interconnects. CBS Interconnects requires that a participating cable system carry four advertiser supported networks, including ESPN, CNN, USA, and MTV. CBS then purchases approximately 30 percent of the availabilities directly from the cable

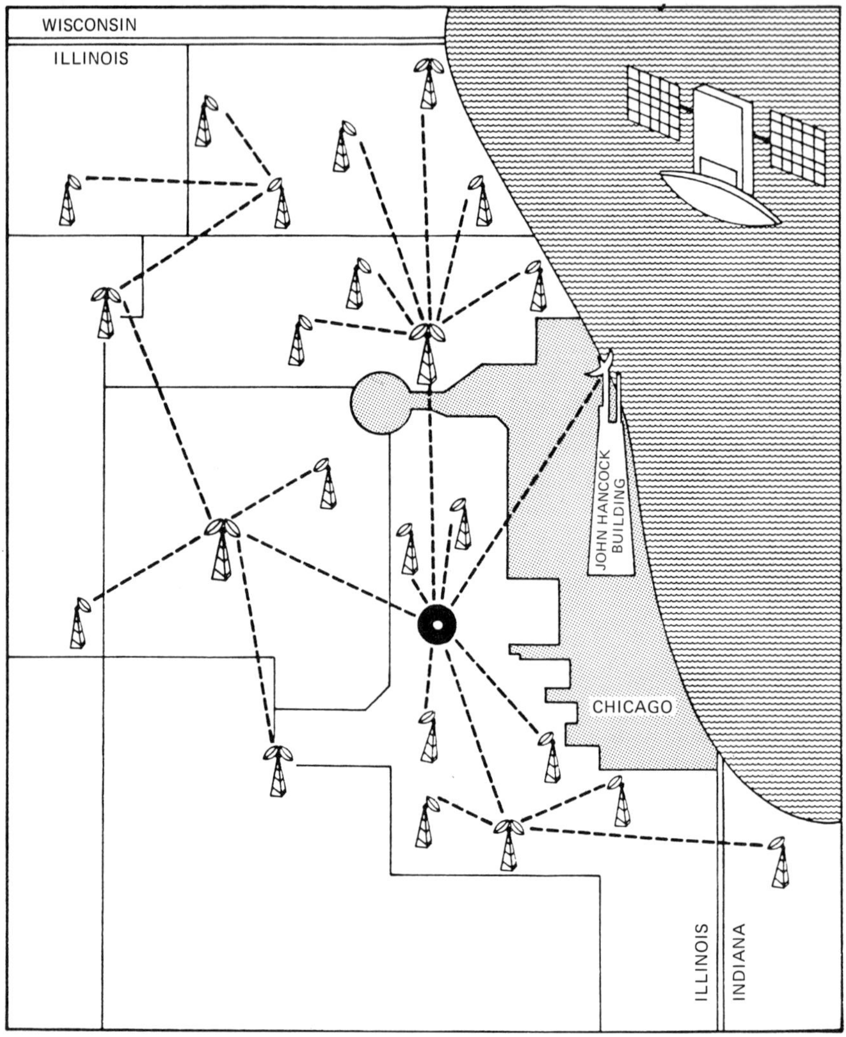

Figure 5-10 Centel Videopath interconnection of Chicago suburbs. (Centel Videopath, Inc.)

operators for resale. This rather ingenious step solves scheduling and traffic problems between the individual cable systems and the interconnect. CBS Interconnect always knows what it has to sell, as do the local cable systems. Some of the Chicago area systems have very productive local sales staffs, while some systems sell no local availabilities at all.

CENTEL VIDEOPATH, INC.
PROGRAMMING LINE-UP

Regional High School Football 1985
Capture the excitement of regional high school football with an eight week series, September 6 through October 26. Potential state qualifying teams have been chosen to generate interest among your community. (Two hours; Producer: Centel Video Productions)

Chicago Spotlight
A local music video program showcasing Chicago area bands and local video producers. Contests, star interviews and give-aways keep time with the music. Reach the young adult crowd. (One-half hour; Producer: Centel Video Productions)

Street Beat
A television magazine featuring concerts, conventions, theaters, restaurants, the elegant, the off-beat and the always exciting events which make the Windy City great! (One-half hour; Producer: Centel Video Productions)

The Beauty Workshop
Top make-up artists demonstrate and advise women on how they can brush on success. A perfect opportunity to reach women of all ages. (One hour; Producer: American Cablesystems)

Body Tunes
A live exercise show featuring tips on diet, health care and supplemental exercise programs. Shape up your media message by reaching health conscious consumers. (One hour; Producer: Group W Cable)

Investments, Etc.
Tax reform, IRS auditing, penny stocks—what do financial issues mean to you and me? A live audience gets the answers from prominent financial guests. (One-half hour; Producer: Centel Video Productions)

Just Between Us
It's strange, it's extraordinary, it's captivating! The Incredible Elizabeth, a proven television personality, uses her psychic powers on prominent guests and events in a live audience interactive format. Daily airings mean sponsors can draw on The Incredible Elizabeth's captivating powers. (One hour; Producer: Centel Video Productions)

Figure 5-11 (Centel Videopath, Inc.)

The CBS Interconnect then places insertion equipment in each of the individual cable system headends, transmits the commercial to the individual headends via microwave, making the switch remotely at the appropriate time. The CBS Interconnect can offer a very attractive opportunity to advertisers who wish to purchase the entire Chicago area.

The interconnect may create its own level of availabilities within the cable network programming and could absorb as much as 75 percent of the local availabilities provided by the network. The remaining spots may be left for the individual cable systems that make up the interconnect to sell independently. In other cases, the interconnect may take over the entire advertising function for all of the member systems and no availabilities will be offered independently in a single system.

An excellent example of local origination programming is again Centel Videopath. Using the production facilities of local cable operators, and joining in production partnership with them, Centel offers a very interesting programming line-up to member operators. The programming line-up is shown in Figure 5-11. Centel uses promotional announcements on the four advertiser supported networks to draw attention to the origination programming. Either the interconnect acts as the sales representative for the origination programming, or the interconnect negotiates an agreement with the operators. For some special programming, for example, an individual cable operator may feel it can generate more revenue by selling the advertising directly. As might be expected, the Centel Videopath Network is growing very rapidly.

Chapter 6

Some Advertising Basics and Cable TV

Marketing plans and advertising strategies begin with a determination of what prospects there are and how they can be reached most efficiently. Copy platforms consider not only to whom you are talking but how you will address them.

More and more cable TV relates to both the *who* and the *how*.

Consider some of the things advertisers and agencies were saying when cable was just a new-fangled way to improve the TV picture in fringe areas. In his 1969 book, *With All Its Faults*, Fairfax M. Cone, head of one of America's largest ad agencies, said:

> . . . the men and women who are the driving force in advertising today should insist that the most dynamic of all advertising media [TV] shall perform in the total public interest. . . . They should support and insist upon a reasonable amount of programming for the minority of viewers whose interests go beyond horse opera and crime shows and the witless and soporific entertainment that is called situation comedy.

A look at today's program guides indicates that most cable subscribers should be able to find something to their tastes. On a given Sunday, for instance, Wolverine Cable of Battle Creek offers congressional hearings on C-SPAN, the public affairs network; sports ranging from Australian Rules football through bass fishing, karate, and rodeo on ESPN, the sports network; fashions on CNN; country music on the Nashville Network; hard rock on MTV; a salute to chanteuse Edith Piaf on the Arts and Entertainment network. The opportunity for highly selective viewers to find what they want exists. Matching appropriate advertising to those viewers is one of today's major media challenges.

Another echo from 1969 is a speech delivered by Barrett F. Welch, entitled, "Getting the Maximum Mileage from Your TV Advertising Budget." Here, in part, is what he had to say:

> We can't always use the standard demographic splits [in finding the target audience] because our audiences are defined not by demographics, but in psychological or physical terms—the over-medicators, the health faddists, the fashion- or beauty-conscious women, and so on.
>
> We need to know more about the kinds of people who buy our product, and we need to spend more time and effort matching these characteristics against our programming. Even though, in the end, our circulation costs may go up, the payoff could be tremendous.

At the same time that Welch, then president of Carter Products, was discussing media, Edward N. Cole, president of General Motors, had this to say about messages:

> We have a new breed of consumer, and only the most intelligent, direct, and personal appeal can hope to gain his attention or earn his purchase. I recognize the importance of advertising being different, exciting, provocative. But common sense and believability are even more basic requirements—particularly when we are selling items which represent a substantial investment of money.

Not a bad precautionary note in relation to today's consumers and the commercials and infomercials addressed to them.

A couple of other points in Mr. Cole's speech seem particularly apropos to the world of cable advertising:

> The consumer has become immune to the big mass sell. It doesn't appeal to him personally, and to a large degree, he just doesn't believe it. . . .
>
> No matter how much emotion may be involved in the purchase decision, the buyer wants and needs rational reasons for making it.

That almost suits the description of what an infomercial is today.

Ronald Kaatz, media expert at J. Walter Thompson ad agency, says, "An infomercial, through the luxury of time, can explain the product's or service's benefits, uses, and characteristics."

Bill Harvey, of the *Media Science Newsletter*, puts it this way: "An infomercial can be regarded as more relevant, absorbing, sincere, and real than a short-form broadcast commercial."

So, with a wide variety of special-interest programming available, the advertiser can narrowcast to just the audience desired. With the 2- to 7-minute infomercial as a tool, the creative person can develop a factual, detailed selling story. Everything yesterday's purchasers and practitioners of advertising asked for seems to be at our fingertips, there for the using.

But cable, for several reasons, is also performing a number of different advertising functions that have evolved in recent years.

Some cable networks (USA, CBN, CNN) are providing audience measurements that indicate how they can beef up the total number of TV viewers available to an advertiser. With broadcast network numbers down, this is an attractive proposition to media buyers seeking an ever-larger total audience at the lowest possible cost. Commercial materials can be the same for both broadcast and cable networks, usually the standard 30-second spot.

Other cable networks have zeroed in on specific audiences: MTV with music for teen-agers and young adults, Lifetime for the health conscious, the ethnically oriented Black Entertainment Network; these and others can deliver a rather sharply defined group of viewers. Many advertisers develop special commercials for each particular audience. Often they stay with the standard 30-second format. Sometimes, however, an entire show is purchased. This gives the advertiser a chance to establish identification through billboards, devise different-length commercials, and integrate show talent into the sales effort. These same opportunities exist now as at the time of the Danny Kaye show, discussed in an earlier chapter. It seems probable that the same rewards in terms of identification and approval are also available. With cable, however, there's the additional advantage of knowing that the kudos will come from a preselected group of people with the right demographics rather than from a mass audience.

So far, at least, cable is relatively free from another basic advertising problem, one that is sure to get worse on broadcast TV: *clutter*. Research published by advertising agency Needham, Harper and Steers indicates that nonprogramming interruptions have more than doubled in the past 14 years. In 1970, the typical one-hour prime-time show would carry 16 commercials, occupying seven minutes and 38 seconds. By 1984, the number was up to 33 and the time had increased to nearly eleven minutes. With less federal regulation and shorter-than-ever spots, it's predicted that by 1990, an hour show will be burdened with 44 different spots using about twelve minutes of time. Paul Harper, chairman of Needham, Harper and Steers, says that by that time, clutter will have become a sort of video smog.

No wonder that, in an attempt to keep their advertising visible and identifiable, some businesses are turning to cable. Procter & Gamble, through Compton Advertising, creates and produces its own show, "At Home with Beverly Nye," for airing on the Christian Broadcasting Network. Through the Benton and Bowles Agency, P&G also produced "The Catlins" for use on cable superstation WTBS.

You can be sure that the commercials within these shows are not surrounded by station promos, public service announcements, and sales pitches from other companies.

Local advertisers, too, have the chance to evade the smog when they utilize cable for advertising. They do, however, face another problem: how to do commercials that compare favorably with slickly produced network spots at an affordable price. The answer may lie more in manner than in money.

Simple, well-thought-out advertising can be far more effective than expensively mounted, gimmicky, brag-and-boast extravaganzas. Credibility and recognition are better criteria than laughter and applause.

The most difficult thing to determine is the essential favorable difference in your product, service, store, dealership, line, or location. Once that has been determined, you might apply Jones's Mirror Theories for retail advertising. They were developed for a National Automotive Dealers Association meeting in New Orleans in the early 1980s:

Theory #1

Your advertising is a mirror. It should reflect you and your dealership [or store] in all its individuality and difference.

Theory #2

Advertising should dramatize, illustrate, and enlarge upon the facts. It's a magnifying glass as well as a mirror.

Theory #3

If you now look like everybody else in your advertising, develop a difference and talk about it. Service? Size? Reputation? Attitude? Location? Search it out. It can be the single most important ingredient in your marketing mix.

Theory #4

Price advertising may be OK and necessary short-term, but it's not for the long haul. If you swing low, somebody else will swing lower.

Theory #5

Don't be irrelevant in your advertising. Always make it reflect your personality.

Theory #6

Get help—an advertising agency, if your budget warrants

it. Otherwise, use your local media people for creative input.

Before we discuss advertising agencies in more detail, here are some additional theories for the advertiser:

Every part of a commercial reflects your taste. Things can look clean and smart without costing a lot.

Computer graphics (which many systems can either provide or can arrange to get) are contemporary and exciting (see Appendix A).

Consumers appreciate significant information, clearly stated, about products, prices, guarantees.

If you are comfortable on camera, you could be your own best spokesman. But to be that, you must be completely organized, rehearsed, and ready. If you're going to do it, make sure you provide time to do it right. You might not be the best judge of your adequacy in this function, so seek some objective opinion.

Use all the help you can get. There are a number of sources that we'll cover briefly.

ADVERTISING AGENCIES

More and more advertising agencies are beginning to give cable TV a place in the media plans they formulate. There has been some resistance in the past, owing to lack of standardized procedures for buying and difficulty of measuring cable audiences. Most cable systems now recognize the importance of selling agency media directors on the proven performance of the medium.

The problem had these dimensions: (1) agencies needed to be persuaded to make a transition from a cost-per-thousand buying mind set, to a qualitative evaluation based on the logic of the programming environment and efficiency, (2) agency personnel were not in cabled areas so they had no personal familiarity with cable programming and function, and (3) much extra work was necessary to make cable buys, involving many copies of tapes and numerous transactions.

This sort of feeling was echoed and re-echoed at systems serving Miami Springs, Seattle-Tacoma, Long Island, Cincinnati, and Columbus. In Tulsa, Oklahoma, one of the most successful systems in terms of developing advertising potential still has a problem:

> . . . overcoming ad agency resistance to cable. Agencies seem to feel that cable is too complicated for them; aren't interested in learning about it, don't know what to ask when they are. Are looking at cable only under pressure from clients.

CABLE AND CREATIVE DEPARTMENTS

Maybe more advertising agencies should get their top *creative* people into the cable arena. Sometimes that's what it takes to find a truly effective media application.

That's why, for years, *Reader's Digest* has had a vice-president and director of creative services working out of New York. He is Anthony L. Antin. Tony knows the top creative people in many, many advertising agencies and works with them to maximize the power of his already powerful medium. Once, at least, the magazine hosted a dozen creative people from the Midwest at its headquarters in Pleasantville, New York. Then Senior Editor John Allen and a group of his fellow editors discussed all manner of subjects with the advertising people, who came away with a clearer understanding of the magazine's possibilities.

Cable networks, systems, and associations need a continuing dialog with admakers. The Cabletelevision Advertising Bureau has, in fact, inaugurated a series of meetings around the country intended to establish such a dialog.

The medium's ultimate destiny could be shaped, in part, by imaginative creative directors, copywriters, art directors, and TV producers, who might welcome freedom from the constraints of 30-second time blocks.

Every sales manager, every salesman for both cable systems and satellite networks might profit from a shared luncheon with an agency creative director. Their reactions could be quite different from the almost predictable reactions of time buyers. Coaxial Cable, of Columbus, takes the mountain to Mohammed if necessary. A large recreational vehicle has been fully equipped as a traveling lunch-and-meeting facility and is on call for presentations and demonstrations in agency parking lots, or anyplace that might be designated by a prospective client.

SEMINARS FOR AGENCIES

Seminars at which all elements of advertising agencies are in attendance could be useful. There's a tendency to limit invitations of management, media, and research people to the "decision makers." This leaves out the important possibility that *effective usage* is very much the bailiwick of "the creatives," those people whose ideas expressed can add exponentially to the value of media dollars expended.

Some systems have already recognized the opportunity. Group W Cable of Santa Monica, California, is now aggressively promoting itself to the Los Angeles advertising community.

"Response has been excellent," the cable people report, "from major agencies, which account for most of the system's sales to date. Agencies recognize both the advantages and necessity of cable advertising, with such factors as declining commercial television network shares, and the fact that cable serves the most upscale communities in the area."

The system is shooting for a bigger share of local advertising and is developing informational programming in collaboration with local colleges and business and civic groups. This includes a series of short programs designed to be "fully sponsored by individual advertisers."

What an opportunity for "the creatives" to prove the power of ideas within a controlled environment where time is flexible enough to accommodate the right execution of those ideas!

Another outstandingly successful system, CPI of Louisville, Kentucky, says this:

> System believes that an important part of its success is its understanding of its product and knowing what to do or not to do. First priority is given to developing programming that will benefit, enlighten, or entertain its audience, *then integrating advertising into the programming.* Knowledge of other media and *being able to talk to agencies in their own language* has also been a valuable asset in generating ad revenue.

AGENCY LEADERSHIP

There are some positive examples on the agency side, too. Campbell-Ewald of Detroit has made a major commitment to the development of cable advertising. In 1983, all agencies within the Interpublic Group of companies were invited to participate in a competition for the Ted Little (a former chairman of Campbell-Ewald, renowned for his innovative ideas) Cable Award. An objective outside jury judged submissions for the best use of cable as a medium, and there were cash awards for the individuals responsible for the winning idea.

This monetary stimulation for agency people attests to the sincerity of Thomas B. Adams, Campbell-Ewald chairman, when he announced the competition. "If we truly want new creative forms [of commercials]," Adams said, "let's invest together in making them happen."

In 1983, Campbell-Ewald also kicked off a five-year, $25,000 college-level competition offering scholarships and summer internships to students. Objective of the competition: "To involve, directly, America's colleges and universities in the new technologies of the cable industry."

Perhaps the best hope of cable TV is the enterprising smaller agency, like J. W. Messner of Grand Rapids. The principals in such companies look upon the wired TV systems as a chance to innovate for their clients. Jim Messner, for instance, sees certain unarguable advantages in cable, as does his media director, Dotty Trommater.

"My oldest client," says Messner, "has been in cable consistently for two very good reasons. They make bedding and box springs and must, naturally, compete with the Posturepedics and Sertas and Beautyrests of the world. They make an excellent product at a reasonable price and should get consideration from the careful buyer. In the Grand Rapids area, we know where those people live, and we can reach them, very reasonably, through several cable systems. Acme bedding has just two stores, they're in town, and we don't need to reach out 50 miles in any direction. So, for costs comparable to local radio, we get local TV."

Another client, a pet-food manufacturer, made a 60-second commercial along with a 30, during a single shoot. Costs for exposing the 60 (which was also used at trade shows and meetings) on regular TV were unaffordable. "However," Jim Messner explains, "we were able

Figure 6-1 Thomas B. Adams, former chairman of Campbell-Ewald, presents a plaque to winners of the first annual Collegiate Cable Competition. The contest, developed by C-E, is to "involve colleges and universities in the use of new technologies in the field of communications." (Courtesy: Campbell-Ewald Co.)

to get good usage on cable, well within the budget, and reached a good number of the people who are our best prospects."

An interesting, if indirect, way of buying cable audience was described by Dotty Trommater: "We wanted to reach audiences in a number of smaller Wisconsin communities, as well as towns in northern Illinois, southwestern Michigan, and western Ohio. By checking dots within a circle on a map, we could pinpoint cable systems that carried a WGN signal. We knew the audience we wanted tended to watch movies on that channel and made the assumption that this would hold true on cable, as well as on direct telecasts. So we bought the show we wanted on the independent TV station and got all the cable systems too."

One cable venture that is on temporary hold at the Messner agency is a show called "Cable Cars." Originally involving about ten car dealers, the half-hour show aired nightly, five days a week, at 7

P.M. on cable Channel 10. Each dealer had a quota of cars to show in detail, with a chance to put a price on them and to invite callers to respond to a designated and displayed phone number. The show was to run for 13 weeks.

"A few of the dealers got discouraged the first night," Messner says regretfully, "and they started backing out before we were really started. They expected more than we could deliver. Other dealers, after a few nights, said that they could see results immediately after the show, but they had to stay open nights to capitalize on them. Economically, these times were rough, too."

The show was canceled, but Messner remains convinced that, with aggressive, promotion-minded dealers, the format can work.

"When the show is known and publicized," he states, "qualified car prospects are sure to tune in and shop. It's the same as classified advertising in the newspaper. Not everybody reads it, but those that do are definitely in the market. We'll bring 'Cable Cars' back one of these days," he promises. "I'm sure of that."

Meanwhile, a kind of programming is beginning to emerge that offers both local and national advertisers a place for compatible commercials.

"Future Sport," produced by the Doyle Dane Bernbach agency, is about athletic performance as measured by computers. Sponsors include Atari, GTE, Mobil, Volkswagen, and Sylvania.

"Working Mother," which deals with the problems of an increasingly large segment of the population, is supervised by Dancer Fitzgerald Sample and has General Mills, Life Savers, Duracell, and Parker Brothers as sponsors.

These shows are tailored for specific audiences that can be reached through the cable networks.

TRADE ASSOCIATIONS

In a comparatively few years, the Cabletelevision Advertising Bureau (CAB) has assumed a tremendously important role in cable's future as an advertising medium. Robert H. Alter, the organization's president, is a veteran of the Radio Advertising Bureau and thoroughly understands the function of an association dedicated to the competitive selling of ephemeral entities like time and space. He con-

Figure 6-2 Robert H. Alter, president, CAB (Cabletelevision Advertising Bureau)

sistently addresses all three segments involved in cable's future: systems people, advertising agency personnel, and representatives of the advertiser community.

For instance, CAB conducts regional seminars for system ad reps. These are hands-on meetings in which the attendees get involved in exercises replicating real situations. Among the subjects covered are the positioning of cable against other media, use of co-op dollars, local cable advertising success stories, and effective cable commercials.

Figure 6-3 (Cabletelevision Advertising Bureau)

At cable advertising conferences around the country, CAB gets the latest and best thinking of top agency people. In Chicago, John Kinsella, chief executive officer of Leo Burnett, indicated a doubling of expenditures for cable by Burnett clients in a single year; in Dallas, Tracy-Locke's chairman, Norman Campbell, warned that cable should be viewed as a new communications medium, not simply as an extension of TV; and in San Francisco, John L. Tormey, executive vice-president of McCann-Erickson, predicted tremendous growth based on the fact that "cable allows copywriters and producers to transcend the boundaries of the 30- and 60-second commercial break. With cable, advertisers can return to the Golden Age of television when programs were fully sponsored by leading corporations."

Advertiser groups, too, get plenty of attention from CAB. Round-table luncheons are held with members of the Association of National Advertisers, with companies like Scott Paper, Gillette, Bristol-Myers, General Foods, and Sears represented by top marketing people.

Among retailers, the cable story as told by CAB is becoming increasingly familiar. Meetings and presentations have involved a carpet seminar; the National Retail Merchants Association; a large shopping-center owner-operator; a fashion group representing major department stores, catalog houses, and specialty boutiques; and a home-center symposium for chain stores involved with tools, lumber, and do-it-yourself merchandise.

A vast amount of material is available to members of the bureau. Clio award-winning cable commercials are on a videotape cassette, as are many other examples of infomercials, co-op spots, and inexpensive retailer messages.

Ad-Tier, an informative monthly newsletter, covers all important and current events relating to cable advertising.

"How-to" brochures on newsletters for systems, on hiring salespeople, on rate-card structures, and on time management are available. A complete directory of organizations representing systems and satellite networks has been completed.

Most interesting are the Cable System Advertising Profiles compiled by CAB. These single-page briefs are packed with information and ideas. Problems in selling advertising are discussed, equipment available for production is catalogued, rates for advertising are set forth, and various systems' differing advertising philosophies are expounded. An examination of this file is a quick look at the state of the art in cable advertising.

Research is ongoing and constant and is devoted to assembling every possible fact, case history, sales technique, or personal story that can attest to cable's effectiveness.

For more information, contact Cabletelevision Advertising Bureau, Inc., 767 Third Avenue, New York, New York 10017, phone (212) 751-7770.

The value of the CAB as a resource was summed up by George Thorry, Continental Cable's regional marketing director in Lansing, Michigan: "The Cabletelevision Bureau and Bob Alter are the best thing that's happened to cable in ten years," he says, succinctly.

Another highly knowledgeable association is the National Cable Television Association, 918 Sixteenth Street N.W., Washington, D.C. 20006, phone (202) 457-6700. Its 2,000 members serve an estimated 70 percent of all cable subscribers. It collects information about cable and audiences that can be of significant value to cable advertising planners.

GENERIC ADS AND NATIONAL SERVICES

It seems inevitable that generic commercials by business category will play a part in the growth of cable TV. Small businesses selling similar goods and services with certain intrinsic values can be served by such commercials. All that's needed, once the basic idea is clearly and memorably stated, is a place for audio-video reference to the dealer's name and location.

To meet this need, Palmer Cablevision of Naples, Florida, produced a reel of ten 30-second commercials showcasing a range of retail businesses and allowing for individual identification. The reel was made available to systems across the country for $250, or $25 a spot. Those interested were invited to contact Palmer Cablevision, 338 8th Street South, Naples, Florida 33940, or to call (813) 263-6500. Bob Michael, Palmer's sales manager, feels that there continues to be a need for such commercials and predicts entry into the business by independent production companies that will keep the spots updated.

"It's probably a field for independent production companies,"

Michael says, "rather than cable channels. There has been a continuing interest in our reels, but they need periodic updating and it's just not our number one priority."

Certain patterns that evolved in radio advertising would seem to make sense for cable. In Detroit, for instance, the Pat Patten Agency specialized in creating radio commercials and buying time for automotive dealers and dealer associations.

During the 1960s and 1970s, Chuck Blore of Hollywood specialized in effective radio commercials for all manner of clients, and Stan Freberg became the king of commercial comedy. Similar specialists can and will evolve pattern TV commercials that can be sold to agencies and clients for use on cable across the country.

CO-OP FUNDS

Every major media association (Bureau of Advertising, American Newspapers Publishers' Association, Radio Advertising Bureau, Television Bureau of Advertising) has developed programs to assist members in obtaining co-operative advertising funds.

No wonder. Recent projections for co-op funds offered by advertisers indicate an unbelievable bonanza of $7.4 billion. The paradox is that, of that amount, nearly $2.6 billion (approximately 35 percent) will go unspent! So it behooves alert retailers and their agencies, as well as all media salespersons, to go after a slice of that giant pie.

Not too long ago, Phil Roberts, who heads The Co-op People (420 Lexington Avenue, New York 10170) said this:

> Retail co-op dollars are not finding their way into cable TV because cable operators don't understand the whys and wherefores of co-op. You can't sell what you don't pitch. Many cable companies don't make any time sales at all because, for the moment, they're more interested in selling HBO to a subscriber.
>
> Nevertheless, cable is soon to become one of the most important avenues for retailer co-op expenditures. Its potential is great.

Certainly the CAB sees the potential. Recently, the bureau completed arrangements with Standard Rate and Data's co-op advertising division to supply member systems with in-depth details about vendor co-op programs. In addition, a Co-op Documentation Form, for adaptation by individual systems, has been developed.

Cable appears to be an ideal way for the retailer to get on TV to promote the brand items he or she carries. As Saralee Hymen, then CAB's marketing VP explained, "Cable is a natural extension of their [the manufacturers'] advertising strategy and is a way to get added exposure for TV commercials they might have produced for use by dealers in local markets."

To further buttress the effort to capture co-op dollars, the bureau informs systems of specific plans that are available for indicated periods of time. Among those covered in one bulletin were Wrangler Division of BlueBell, Inc., Norelco, Parke-Davis (for specified products), and Searle for itemized over-the-counter drugs. Also covered were plans offered by Union Carbide, Shell Oil, Mobil, and Pennzoil.

The size of the co-op opportunity can be partially measured by the fact that more than 600 national advertisers work with the Advertising Checking Bureau on co-op plans. That organization is responsible for verifying media rates charged and has asked for rate cards from all cable networks and systems.

Many local systems are starting to pay more attention to co-op funds as a source of advertising revenue. Cox Cable of San Diego added an account executive to work exclusively on co-op possibilities for local clients.

In Lansing, Continental Cablevision has joined BACON, the Broadcast Advertising Co-op Network, and gets a continuous update with full details on active programs. For the most part, it is the responsibility of cable systems to point out to advertisers the existence and value of already available co-op programs, particularly for retailers with a broad range of branded merchandise.

Actual forms such as a Dealer Request for advertising materials (see Figure 6-4) are made available to local retailers by Continental's sales staff. In *Connections*, the organization's newsletter, a "Co-op Corner" has been established so that retailers in their service area know what's new in the way of available funds. These are the kind of briefs that are included:

> ADIDAS offers co-op support for all first-quality, regular-priced Adidas shoes, clothing, and accessories and encourages pooled programs [several retailers underwriting a specific program]. Participation

CHRYSLER
CORPORATION

DEALER REQUEST FOR "PICKUP CASH REBATE" PROGRAM ADVERTISING MATERIALS

PLEASE TYPE OR PRINT FIRMLY

REQUESTING DEALERSHIP

Dealership Name

Zone Dealer Code

( ) Dealership Phone Number

Authorized Dealer Signature

MAIL ALL COPIES TO ZONE OFFICE FOR APPROVAL

ADVERTISING MATERIALS REQUESTED (Tapes: Air Quality—Ready for Dealer Tagging)

CHECK APPROPRIATE BOX(ES)

☐ Dodge Full-Line Pickup Rebate Radio Commercial

☐ Ram 50/Power Ram 50 TV Commercial

☐ Full-Line Rebate TV Commercial

☐ Newspaper Ad Slick
Ship Direct to Dealership

SEND MATERIALS TO

RADIO STATION(S)

TV STATION(S)

☐ Requested deviations are attached. I realize that Zone Manager's approval is required.

ZONE AUTHORIZATION This section is to be completed by your Zone.

Dealer's pickup build out order status has been verified and he is qualified for:

☐ **ELIGIBLE DOMESTIC PICKUP ORDERS** × **$50.00** = **$**

Requested materials to be sent at once to stations shown.

Zone Truck Sales Manager Date

If requested, advertising deviation has been reviewed and . . .

Approved ☐

Disapproved ☐

Zone Sales Manager Date

DISTRIBUTE COPIES AS FOLLOWS:

1. ***White Copy***—*Forward to Program Headquarters (After Zone Approval)*
2. ***Canary Copy***—*Send to local K&E office for forwarding of advertising materials to the dealer or to selected media locations*
3. ***Pink Copy***—*Return to dealer to be used for CO-OP advertising fund payment request. Attach invoices, tear sheets, etc. to Program Headquarters, P.O. Box 1110, Birmingham, Michigan 48012*
4. ***Goldenrod Copy***—*Return to dealer for his file*

Figure 6-4 Advertising Allowance Request (Courtesy of Chrysler Corporation)

of 50/50, up to 1 percent of net purchase within a calendar year, and locally produced commercials are acceptable.

NORITAKE offers 100 percent co-op on 10 percent of net purchases for advertising of Noritake imported china products. Noritake encourages pooled programs and reimburses each cooperating retailer by check. Nationally produced television commercials are available.

CHAMPION offers 100 percent co-op to retailers when manufacturer-prepared commercials are used, and 50 percent participation when retailer/wholesaler-prepared commercials are used.

If advertisers—particularly retailers—cable systems, and local agencies would get their heads and their plans together, that 35 percent of unused co-op money might not be idle.

A little leadership could develop "pools." Several retailers who handle the same branded merchandise might join in advertising its availability at their places of business. Leadership by distributors can bring jobbers together to take advantage of brand co-op money.

Leadership by cable systems in their business communities can make new money for advertising available and brighten the local retailing picture.

THE CABLE SYSTEM

The single greatest tool available to the local advertiser who is seeking an understanding of cable TV is the local system. If the community has cable service but there are no advertising availabilities, interest can be indicated to the system's management. Enough interest might move the cable operators toward decisions that will improve the situation.

If, on the other hand, the system is geared for advertising, it will have an advertising sales representative. Like any good salesperson, the cable rep *knows the product* and knows what is available, be it on satellite networks or of local origination. Previous experience with clients can be useful in planning such things as frequency and duration of a campaign. Examples are available of cable commercials that have done the job for other advertisers.

Usually the rep will have some knowledge of copy and production. Larger systems will have additional people who double as copywriters and scripters for local shows. They will be made available to help in transforming sales thoughts into workable, affordable cable programs.

Quite often the system's producers and production facilities will come with the contract and be available for commercial making.

It will be very much a team effort, with the advertiser and the system's trained personnel exchanging ideas until a strategy, copy approach, and production technique are matters of mutual agreement.

Many cable systems offer to make commercials free of cost for the client, and sometimes the initial commercial is free. Cable operators will give clients a firm cost estimate before production. Sometimes systems without the necessary equipment work with local production houses at very reasonable costs.

Commercial costs (as of mid-1984) are quoted by various systems as follows:

Continental of Ohio
On-location spots \$75, in-studio \$60, slide spots \$45. (Rates will be raised in the near future.)

Village Cable (Chapel Hill, North Carolina)
Typical 30-sec. spots \$100–\$175, half-hour program length infomercial \$150–\$350.

Tele-Cable Sales (North Syracuse, New York)
Commercial production is subcontracted out to several area production firms. Average cost per 30-second spot is \$300.

Albuquerque Cable
Local 30-second ad using slides costs approximately \$250 to produce, including copywriting, audio talent, audio and video studio time, and photography.

Comcast Cablevision (Willow Grove, Pennsylvania)
First commercial is free. System then charges \$250. Additional spots cost \$100 if produced at the same time.

Group W Cable (Santa Monica, California)
Production charges negotiable, with most 30s being charged at \$800 to \$1,000.

Warner Amex (Cincinnati, Ohio)
\$200 per hour for full studio facilities, including two cameras and staff, \$75/hr. editing time.

Even the highest of these charges are incredibly low to those accustomed to the kinds of costs incurred in making 30-second network TV spots, where a $40,000 price tag isn't considered exorbitant. The advertiser must insist, however, that the cable commercials hold up with the $40,000 spots in the same television environment. To this end, they should be carefully planned, kept simple and direct, and done with a minimum of on-location fuss.

Appendix B presents a detailed plan for creating a cable television commercial.

Chapter 7

Commercial Flexibility

Sponsorship, as illustrated by the examples already discussed, allows control of commercial length within the sponsored time segment. It permits the kind of innovative selling messages referred to earlier in conjunction with the Danny Kaye show and Lowell Thomas's "High Adventure." It provides for the kind of integration that retains viewer interest and involvement. More than anything, it offers the creator of commercials *time*. This same element, up to 120 seconds' worth, is available to advertisers who buy into some cable network programming. In local origination channels, there is no time limit at all.

THE INFOMERCIAL

The result, in current parlance, is the *infomercial*, a commercial message that takes more than 60 seconds to provide a depth of information not possible in the usual 30-second and occasional one-minute commercial.

This flexibility is being put to good use by some advertisers. The Fisher-Price Toy Company, for instance, devised some informative and interesting messages in which the people who design and make toy helicopters, trucks, and dolls talk about the care that goes into crafting safe, dependable, and amusing playthings for kids. These 90-second infomercials appeared on cable shows aimed at children.

Another advertiser interested in homemakers is Corning. Its cookware is demonstrated through long-form messages featuring an affable chef, Cornelius, and his interlocutor, Gene. They spend three minutes with a recipe for "nesteggs" during which four different Corning items are displayed and used. In a five-minute segment, the same talent explores the advantages of using Corningware in microwave cooking and offers a booklet on the subject to anyone who will write for it.

Some advertisers (see Figure 7-1) use the extra time to create mood rather than sell features:

> Kawasaki stages a two-minute motorcycle ride that employs split screens, solarization, slow motion, and breathtaking photography to give the viewer the vicarious thrills of trailbiking.
>
> Lincoln-Mercury develops an outrageously funny staging of the Capri that uses hula dancers, Japanese tourists, motorcy-

Figure 7-1 Storyboard for two-minute Dodge Daytona Turbo Z commercial. The car drives off the billboard and comes to life. (Courtesy: Chrysler Corporation)

> cle policemen, crashing surf, and a bewildered beachwalker to focus attention on a car that is "pure magic." It makes a fast and unforgettable minute and a half.
>
> Black and Decker does everything but knock on your door with an engaging salesman-to-homemaker demonstration of its Dustbuster. It's a micro-vacuum cleaner that shows its abilities at picking up dust, kitty litter, and approval in a three-minute time frame.

An interesting opportunity for local advertisers to give viewers something extra was devised by CNN in conjunction with the 1984 Olympics. Seventy-five spots, each a minute and 45 seconds in length and telling the story of one Olympic athlete, were produced. Local advertisers could buy the series, add 15 seconds of local identification, and have a two-minute spot about effort and excellence. Jack Dykstra, an enterprising Lansing car dealer, considered buying the package to give institutional overtones to his usual hard-sell messages.

Although "infomercial," "advetorial," and "promercial" are words coined by the cable generation, the idea of long, integrated commercials has a rather glorious history.

In a book entitled *Television's Classic Commercials—The Golden Years,* author Lincoln Diamant presents 69 examples of the best in video selling between 1948 and 1958. Slightly more than 20 percent of these commercials were more than 90 seconds in length. One, for Westinghouse, ran seven minutes and was done in two parts. It was a definitive demonstration of the superiority of a washing-machine filter system. Most of the long commercials concerned automobiles, automotive accessories, and appliances. But even Lipton's soup was good for a four-minute demonstration and recommendation from super TV salesman Arthur Godfrey. He used it to sell a relatively new concept, dried soup, against the entrenched concept of canned soups.

Whatever the advertising strategy is, it's advantageous to have time to properly execute it. Chevrolet, for instance, was struggling to get public acceptance for avant-garde styling in 1959. So, using a fine French film company, Arco, and the incomparable sights of Paris, Chevy's advertising agency, Campbell-Ewald, staged a two-minute tour of the fashion capital of the world (Figure 7-2). A driverless Impala promenaded the streets of the city, winning Gallic glances of

Figure 7-2 Longer commercials, offering unusual creative opportunities, are possible through some cable buys. This two-minute winner at the Cannes Film Festival, "An American Visits Paris," dramatized styling for Chevrolet. (Courtesy: Campbell-Ewald Co.)

approval on its way. John Crosby, the TV critic, responded to this 120-second effort by writing:

> The Chevrolet commercial showed a Chevy wandering through Paris down the left bank, around the Arc de Triomphe and down side streets where flower vendors poked their heads through the window. It's one of the best and has the simplicity of great sophistication.

The commercial went on to win first prize for live-action films two minutes or more in length at the Cannes Film Festival.

It's not the awards that are important, it's the involvement. Neither Nielsen nor Arbitron tell us much about that. But we do know that identification, which is certainly related to involvement, has been and is a problem.

Harry Wayne McMahan, columnist for *Advertising Age* and probably the man who has written most about TV selling, had this to say in the mid-seventies:

> For TV, the worst of news is that of the 16 of every 100 viewers tuned to a program who remember a commercial, only eight correctly identify the sponsor. The other half include four who "don't know" and four who misidentify, crediting that commercial to a rival sponsor.

The obvious reaction is to "fix" the commercial: mention the advertiser's name twice as often and keep the logo on the screen longer. Traditional wisdom, as usual, fails to deal with the real problem, which is simply mental and physical tuneout engendered by the unreasonable assault on the senses made by today's commercial TV methodology.

The baby, viewer involvement, has been thrown out with the bathwater, sole sponsorship and random-length commercials.

IDENTIFICATION OF ADVERTISING AND PROGRAM

Cable, for a number of advertisers mentioned earlier, is offering the chance to bring the baby back; to regain the probably priceless ingredient of unmistakable association of the product and the sponsor with the advertising that prospects see.

Ownership of a show, with identifying billboards and closings, is possible for local merchants. "World of Photography" is a network quality production featuring famed photographers, like Dennis Brack of *Time* magazine's Washington bureau, plus segments on night photography, underwater shooting, and so forth. In Lansing, its exclusive sponsor on Continental Cable is Linn's Camera Shops. Locally made commercials are inserted, along with manufacturer-supplied spots for Kodak film and Olympus cameras. In other cities, other camera shops assume sponsorship and build a franchise with photo enthusiasts. There are many special-interest shows available for advertisers seeking to identify themselves with special audiences.

In a sense, cable offers the video advertiser the same sort of opportunities that the universe of special-interest magazines offers in the print area. Just as one can find the right-size ad for the budget in a magazine that reaches real prospects most efficiently, so cable com-

Figure 7-3 Cable infomercial, advertised in *TV Guide* as a program. "Shortcuts" is a trademark of General Foods Corporation. (Advertisement reproduced with permission of General Foods Corporation.)

mercials of various lengths can inform and persuade selected audiences.

Obviously, the selection and purchase of print media became more complex as mass magazines faded and interest-directed magazines burgeoned. Today, video time buying is becoming more demanding as cable offers new choices. The more options offered, the more thought and effort are required for selection. Commercial flexibility means additional responsibility for advertising managers and for advertising agency media and creative directors. For today's writers and producers, it means creative opportunities that simply weren't available in the restricted 30-second format.

It should be pointed out that flexibility, not just greater length, is what cable brings back to television commercials. Often the traditional 30-second spot is considered adequate. In a research project conducted by Richard F. Beltramini of Arizona State University, there was little evidence that advertising professionals wanted more room for prime time TV messages. The optimum length for commercials indicated by respondents was just under 40 seconds, with 65 seconds suggested as a maximum. Obviously, results were weighted by the familiar forms. Many respondents were unfamiliar with the concept of infomercials; in fact, 53 percent of the agency creative directors involved didn't really understand the term.

Among Beltramini's conclusions are these:

> Creative directors' relatively lesser receptivity [to infomercials] will require further modification as the challenge accelerates to design commercials to achieve awareness and understanding through viewer involvement in details of the product or service's function. As advertisers begin to experiment with infomercials, agencies need to adjust their thinking to be ready to act rather than react.

POLITICAL ADVERTISING

Politicians are now discovering cable advertising for many of the same reasons businesses are, and for some special ones. Research reports have indicated that 76 percent of the cable subscribers voted in 1980, versus 62 percent of all registered voters. Cable subscribers were also much more likely to write to an elected official, to make a political contribution, to work for a political candidate, and to work on fund raising than were nonsubscribers.

Cable allows the political candidate to go directly to the right audience on particular issues (for example, abortion and women's rights on the Lifetime network) and to specific ethnic groups (Black Entertainment Television, Spanish International Network, National Jewish Television). Furthermore, cable can match up much better than broadcast television with voting wards, precincts, congressional districts, and so on. And the TV 30-second spot, so often criticized as uninformative, can give way to infomercials.

A U.S. congressman, Barney Frank of Massachusetts, credits cable with his 1982 victory. Polls showed that Frank, a Democrat, had to win Fall River, a Republican town with a 50 percent cable penetration. Three groups in Fall River were targeted with cable programming: residents of Portuguese extraction, veterans and senior citizens, and people in the fishing industry. A "Meet the Press" format was created, with the questioners on each program representing the three targeted constituencies. The three half-hour programs were presented on cable at $150 each. Thirty-second extracts were inserted in the local availabilities on cable networks. Volunteers were used to distribute leaflets promoting the cable appearances. John Florescu, president of Video Productions International, who was Frank's media adviser, says, "Cable democratizes the use of television. Until now, an appearance on TV was a privilege reserved for multimillion-dollar campaigns. Cable enfranchises a new class of office-seeker, right on down to the water commissioner."

ZAPPING

A new fear among advertisers is that cable, with its remote-control units, will permit the "zapping" of commercials. This means that the cable subscriber switches to other material during a commercial. (The broadcast audience cannot do this during the parts of the day when the networks are broadcasting, because the commercials come on at the same times on all three networks.) Therefore, the only way to reach the cable audience that zaps commercials on broadcast channels is to advertise on cable. The cable commercials may also be zapped, but it is theorized that if the commercial content is compatible with the program, which is more likely in the targeted programming and advertising on cable, then the commercial is less likely to be zapped.

Chapter 8

Evaluating Cable Advertising

There is only one hard-nosed way to evaluate radio and television advertising for goods and services, and that is in *sales* that can be related directly to the advertising. Unfortunately, we seldom have such a direct means. The most common substitute is to count the number of people exposed to the advertising. At least if people are exposed to the message, it has a chance to do its work. Further, it is possible to compare one advertising medium to another in terms of the sizes of audience reached, and then to price advertising time accordingly.

Broadcast advertising audiences are measured by meters, telephone surveys, and diaries. Because of the high cost of such techniques, and because of difficulties in getting people in the sample to be conscientious in reporting their viewing or listening, these methods are actually only very crude estimates of audiences. But in the absence of better information and actual sales-effect data, they are heavily relied upon by advertisers and advertising agencies.

NATIONAL AUDIENCE DATA

National audience data are available for advertiser-supported cable networks. Nielsen includes any cable network, that has more than 15% of U.S. TV households subscribing, in national ratings. These figures, very useful to cable network advertisers who are buying time on the whole network, can be used to estimate the audience for a local cable system (see procedure described below). Arbitron and Nielsen include the cable audience in some cities in their regular surveys of individual broadcast markets. However, these data are the cable audience within the ADI (Area of Dominant Influence) or the DMA (Designated Market Area). These are the trade terms that mean an entire broadcast television market (about 70 miles in radius from the city center), so the audience for a program within a cable system that is only a small part of such a market is not discernible.

THE PROBLEM OF LOCAL DATA

Audience counts for *local* cable systems, in their own service area, are very difficult to obtain, for a number of reasons.

Cable systems often serve small areas, and because sample sizes must be the same for small areas as well as large, it is relatively expensive to survey each area exclusively. Unlike broadcasters that share costs of surveying a common market, a cable system has no help. It serves an exclusive territory. Total advertising revenues for a cable system are not, at this point, very significant; therefore, the cost of regular surveys to be used for advertising sales would be prohibitive. Furthermore, there are methodological difficulties in measuring cable audiences. With 30 or more channels available, huge samples would be necessary to break out audiences for each channel hour by hour or program by program. Also, because the cable viewer uses many more television channels than do broadcast viewers, the recall of viewing is much less reliable.

We can conclude from these facts that local cable systems will not have regular, dependable audience surveys from their own service areas in the near future. This has deterred many national advertisers and large advertising agencies from the use of cable. It has been difficult for them to find means of assessing the merit of cable advertising in specific markets.

As noted, national and major market ratings are available for some cable channels, such as WTBS, USA, ESPN, CNN, and MTV. These are helpful to the traditional media buyer but can be deceptive in terms of decision making. They tend to place network cable advertising in the same category as broadcast radio and television.

But the problem remains for local cable, for which there is no measure. Buying cable advertising is much like buying newspaper advertising. The advertiser knows the number of subscribers (the circulation) but has no way of knowing whether or not those subscribers are exposed to a particular message. And buying cable is also like buying small-market radio stations that cannot afford regular, if any, ratings surveys. Nonetheless, advertisers have been effectively buying both newspaper advertising space and small-market radio time from the origins of those media without having indexes of advertising exposure. Prudent advertising in these media demands careful attention to other factors. This is generally referred to as qualitative buying—relying on good judgment and purchase data. Local cable advertising, when considered as a medium, must often depend on case histories as the best evidence of effectiveness.

GETTING BY WITHOUT NUMBERS

One means of informally tracking cable advertising is to isolate advertising for the cable medium—that is, not to advertise a particular product, service, special, or sale *in any other medium than cable.* Any improvement in business or store traffic, or inquiries for the cable-advertised product or promotion, can then be attributed to cable.

A case in point is Morgan's Jewelers of Michigan. Morgan's has used cable to sell one or two products at a time ("ESPN has sold more Rolex and Omega watches for us than I could tell you.") With this emphasis, it has been easy to relate sales to advertising of those products on cable. Tom Gatti of Tele-Cable Sales, a cable rep firm, tells of a department store in Syracuse that started its Columbus Day sale of a particular item ten days early, using local availabilities on cable's USA network. The store's newspaper campaign for the sale did not start until the actual day before the sale. More than half the sales volume was a result of the cable ads. He also mentions restaurants giving a free carafe of wine or a $5 credit toward the check, and ski resorts discounting lift tickets, when cable ads are mentioned. Counting the requests for carafes, credits, or discounts tells the story, pro or con.

Some very large retailers who are heavy in newspaper, radio, and television advertising could not afford to use cable alone to test its value. In these circumstances, it is still possible to use cable only for promoting a single product or special. Increased sales of the cable special are then clearly attributable to the cable medium. This technique might be called "electronic couponing." The cable subscribers are told to look for or ask for the cable special in stores.

This concept is carried out to its fullest through the "Cable Shopper Card" service of Coaxial Communications in Columbus, Ohio, which was described in Chapter 3. People who come into retail establishments seeking merchandise advertised under the Cable Shopper program and display their Cable Shopper Card (Figure 8-1) have been exposed to the cable advertising or have talked to someone who has. Whatever the case, Cable Saver Club sales are "tracked" to cable.

Some sales, of course, are made directly by cable. When the advertiser offers the product for order by 800 phone number or mail,

Figure 8-1 Cable Saver Club wallet card entitles holder to merchandise discounts. (Coaxial Communications, Columbus, Ohio)

a sale or inquiry can be traced directly to the cable system running the advertisement.

These direct-response ads aren't all for *Slim Whitman's Greatest Hits;* they include investments, cookware, and subscriptions to *National Geographic, Newsweek, TV Guide, Golf Digest,* and *Sporting News,* as well as big-ticket items such as cordless phones with electronic memory. Cable attracts direct-response advertisers for two of the same reasons that it attracts any other advertisers—upscale audiences and targeting. But, also very important, cable networks have enough available inventory to accept per-inquiry advertising in prime time. Per-inquiry advertisers are generally relegated to the fringes of network broadcast time.

Reaching for the cable upscale audiences has upgraded the advertisers and the quality of commercial production in direct-response advertising. The general trend toward direct-response marketing, the fastest-growing marketing area, has brought more companies into the field. Consumers are being educated to trust direct-response purchasing. So far, few local retailers have entered the market, but the combination of cable availabilities and trusted retail establishment names could be very successful. The door is open now, particularly in these days of plentiful local availabilities.

USING NATIONAL RATINGS

As indicated earlier, local cable systems are not surveyed regularly to determine audience size by channel and program. However, some of the national cable networks are measured by ratings services. By making broad assumptions, these figures can be used to estimate the local audience. If, for example, USA network has a 2.5 rating nationally for a particular time period, it can be assumed that it has a 2.5 rating locally.

Since the rating is a percentage of all television households in the United States viewing the channel in the time period, it cannot be applied directly to cable subscribers. In 1985, about 50 percent of U.S. households subscribed to cable. By dividing .50 into our rating of 2.5, we obtain 5.0, which is the rating of the channel among cable subscribers—2.5 percent of U.S. households were watching the USA network, 5 percent of the cable subscribers. Dividing the rating by the cable penetration percent of the United States will always give the rating among the cable subscribers. That penetration, in 1985 about 50 percent, will go up a little each year, so it has to be updated periodically with the increase in cable subscribers. The cable rating, rather than the U.S. television household rating for a cable network as reported by Nielsen, should be applied to the number of basic cable subscribers. (*Note:* Sometimes a cable system puts an advertiser-supported cable network on a second tier of cable service, called "extended" or "expanded" basic, or some other, more attractive term. An additional monthly fee must be paid to obtain channels on this tier. For estimating audiences in this tier, which will have fewer subscribers than the basic service, it is necessary to know the number of subscribers in the tier.)

A cable system that has 50,000 basic subscribers, using our national rating of 2.5 for the USA network, will have:

$$\frac{.025 \text{ (national rating)}}{.50 \text{ (national cable penetration)}} = .05 \text{ (5\% cable audience rating)}$$

$$.05 \text{ (cable audience rating)} \times 50{,}000 \text{ basic subs} = 2{,}500 \text{ viewing households}$$

It must be remembered in using the national figures that the daypart to which the figure applies must be the same as the daypart in which an advertiser is interested locally. A 2.5 percent rating in prime time does not mean that that kind of rating would be obtained in an early-morning hour.

There may also be some demographic information on a national cable network. Again, it is reasonable to assume that these demographics will be similar locally. If the program or network attracts high-income males nationally, it probably attracts high-income males locally. This assumption may be somewhat less valid if the demographics of a community are quite different from the national norm.

The cable system sales representative should have recent rating books available, or reprints of the data from professional journals or newsletters. An advertising agency would also know these figures.

ESTIMATING LOCAL CABLE AUDIENCE

Since we know roughly how many television households are viewing television during the various parts of the day, and we have a rough idea of how cable audiences divide their time among viewing options, it is possible to get an idea of the local audience for cable networks without having national figures. If one is conservative, audiences arrived at in this way may give the advertiser a means of evaluating the price of local cable network availabilities. An example follows.

The table below indicates the percentage of television households likely to be viewing television at various times during the day. These are based on national figures representing all television households (cable and broadcast). We can consider them conservative, since cable households are likely to be somewhat higher in overall television viewing than all U.S. TV households.

ESTIMATES OF TV USAGE

7–10 A.M.	10 A.M.–1 P.M.	1–4:30 P.M.	4:30–7:30 P.M.	Mon.–Sun. 8–11 P.M.	Mon.–Fri. 11:30 P.M.–1 A.M.
17%	20%	26%	41%	57%	28%

Source: A. C. Nielsen Company, May 1982. (May is selected, to be conservative. It is one of the lowest viewing months, excluding summer.)

We can assume, conservatively, that cable households will be viewing at least at the level of all television households. In using our cable system with 50,000 households, the expectation is that 57 percent of the 50,000—28,500—are viewing in prime time between 8 and 11 P.M.

We know from audience research that the average cable household spends about 13 percent of its time with the major advertiser-supported cable networks. The rest of the time is devoted to pay TV channels, broadcast stations, superstations, and other programs in which the cable system is not selling advertising. If our example cable system carries USA, ESPN, MTV, CNN, CBN, and Lifetime, roughly 13 percent of the viewing would be to these channels—in our example, 13 percent of 28,500 households viewing 8–11 P.M. This is 3,705 households. If the local cable advertising buy is *all* the cable networks on the system, then that audience could be expected. If the advertiser is buying only one of the six channels, roughly one-sixth of the 3,705 households, or 617 households, could be expected.

If you think cost per thousand (CPM) should be about $11 for television, then you should be paying 617/1,000 × $11 = $6.79 per spot. The cost per thousand homes in cable should be about the same in broadcast radio or TV, although the advertiser may be willing to pay a premium for cable audiences where demographics are particularly suitable for the product or service.

This method of calculating a cable audience for advertising messages and CPM is crude, but it does give a reasonable approximation of the magnitude of the expected audience. It can help the advertiser judge the value of the audience in relation to its price.

MONITORING VIEWING

Two-way cable systems are capable of monitoring the viewing of all cable channels at whatever intervals are desired. For those few systems that are operationally two-way throughout the subscriber network, this monitoring would constitute a census of subscriber viewing. This would be, by far, the most reliable measure of audience for any advertising medium.

However, most cable systems are only two-way-capable, meaning that return amplifier modules must be inserted in trunk and feeder lines, the system must be "deingressed," and certain headend and household hardware must be installed. It is possible to make all these adaptations for only a sample of the system at minimal cost and without making the entire system two-way. The cost would be about $50,000. This is a reasonable cost for larger cable systems and *is an essential step in full development of the cable advertising business.* The data obtained from such a monitoring system would be invaluable to cable programmers, cable advertising sales staffs, and, of course, the advertisers and advertising agencies.

CABLE SYSTEM/ADVERTISER RESEARCH

Advertisers and cable operators may, on occasion, want to conduct local research to estimate audiences. A simple set of procedures is provided in Appendix C.

Appendix A

Computer-Generated Graphics

Computer-generated graphics refers to the images that can be displayed on a television screen resulting from the output of a computer. This is a relatively new option for television production, which previously had to rely entirely on the output of the television camera to produce the images on the television screen. Using traditional television production techniques requires that all the visual images be created outside the production hardware and "photographed" by a television camera. By the use of computer-generated techniques, the images can be created and manipulated inside the hardware using the computational power of the computer. The computer-generated techniques, when they can be used, usually result in a considerable cost saving over similar production using traditional techniques. The computer-generated techniques also allow new kinds of visual images that would be extremely difficult to create using the traditional techniques. Computer-generated techniques are becoming an increasingly important part of television production and will no doubt be an even more important part in the future.

VIDEO TECHNOLOGY

To comprehend how computer-generated techniques can help produce high-quality, low-cost television commercials for cable advertisers requires understanding of a few things about how television works. The image that appears on the television screen is actually a series of still images that change frequently, approximately 30 times per second, which creates the illusion of motion. Each still frame consists of a series of small dots, which can have different intensities of "gray" or of any of the three primary colors. The smaller the individual dots, the less visible they become to the human eye, resulting in a better image on the television screen. The larger the individual dots, the more visible they become, causing a poorer image on the television screen. Hence, the relative size of the dots that create the image on the television screen defines the *resolution* of the image. The smaller the dots, the higher the resolution, meaning that the television picture will be sharper and show more small detail.

In order for television to become a mass medium, a resolution "standard" was needed so that all television sets could tune the same kind of signal. Currently the U.S. standard is 525 dots per inch. Higher-resolution television would mean more dots per inch.

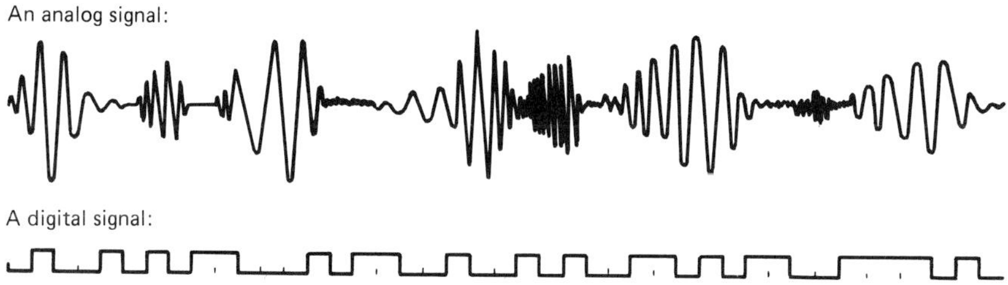

Figure A-1 Difference between analog and digital signal

Television signals are transmitted in "analog" form, meaning that the signal is a continuous wave, with the picture information contained in changes in the waves. This can be contrasted with "digital" transmission, which is a sequence of "ons and offs," or pulses. Figure A-1 shows the difference between an analog and a digital signal.

Most of the existing telecommunication systems today use analog signals, including both telephone and television. If these systems were being built today, they would probably be built as digital systems. It is relatively easy to convert an analog signal to a digital signal, and vice versa. The process of converting a digital signal to an analog signal is known as modulation, and the reverse is called demodulation. Hence, the device that makes possible digital communication over analog telephone lines is called a "modem," or *mo*dulator-*dem*odulator.

The advantage of a digital signal is that it can be directly processed by a computer. This means that the digital codes, the patterns of "ons and offs," can be easily manipulated or changed. Manipulating the digital pattern requires a series of instructions that the computer can use to make the appropriate changes. The series of instructions are usually called programs, or "software." The physical devices that make up a computer are called "hardware."

The special hardware that is needed to create graphic images on a television screen must be able to make the conversion between the digital codes internal to a computer and the video signal that appears on the television screen. This device is called an "interface." The dots that appear on the screen are termed "pixels" and represent the smallest area of the screen that can be changed by the system (Figure A-2). Again, the smaller the pixels, the better the resolution, and consequently the higher quality the image.

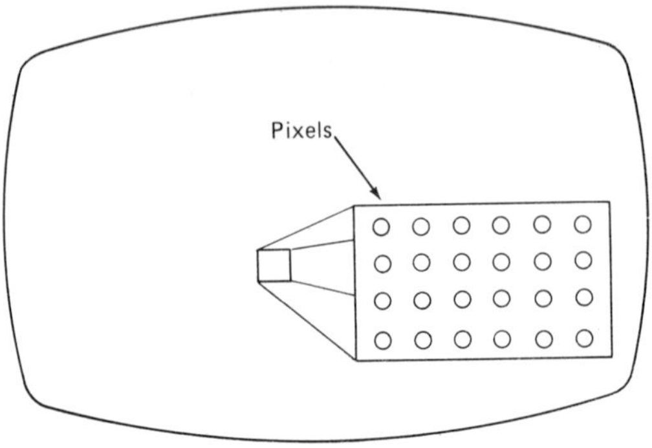

Figure A-2 TV screen

Most graphic interfaces have color capability, although the sophistication and number of colors vary. Today the color that is generated is much like the spot color that appears in the Sunday comics, with areas of single colors. Individual colors are not well mixed and do not provide the entire spectrum. However, the sophistication of these interfaces is rapidly increasing, and no doubt future devices will be able to reproduce images comparable to those created through a camera.

The sophistication of the software used to generate the images can make a substantial difference in the quality of the final product. Obviously, more sophisticated software allows more different kinds of manipulations of the images, as well as shortcuts. Probably the greatest limiting factor in the ultimate quality of the image is the software. Actually, the image itself is a series of computer instructions or software. Along with dramatic improvements in the hardware in the future, there will be even more dramatic improvements in the available software.

CREATING COMPUTER-GENERATED IMAGES

An image can be created for a television screen in essentially three ways. The first is digitalization of a video image created by a television camera. This allows the use of any artwork, photographs, or videotapes as a starting point. The computer can then make changes or additions to the image. Usually high-resolution cameras would be used for this purpose.

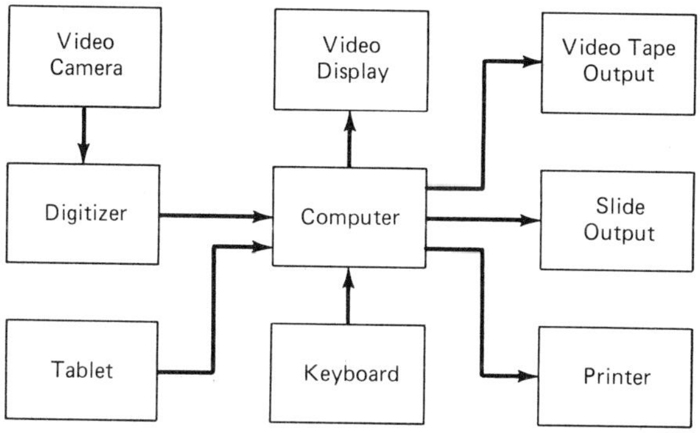

Figure A-3 Typical graphics computer system

A second way images can be created is through the use of a tablet or similar device. A tablet consists of a surface like a small desktop and a special pen that allows the making of drawings much like traditional pen-and-paper sketches. Of course, work done on a tablet could be easily combined with work from a digitalized picture. A typical system, with the digitizing camera, tablet, and computer system is shown in Figure A-3.

The third way of creating images is through the use of commands available directly through the graphics software. The commands would be instructions normally entered through a keyboard. They would allow drawing geometric shapes, rotating images, inserting text of different sizes and fonts, adding and changing colors, and editing pictures. The images can be animated—that is, portions of them can be made to move. Animation requires addition of timing information, sometimes referred to as "blinks," and can be quite sophisticated. Many different elements in the image can move simultaneously, creating very realistic and exciting animation.

As the area of computer-generated graphics continues to develop and improve, no doubt much more production will be done using the computer techniques. Computer-aided graphics are generally much less expensive than traditional methods and can be done in much less time. Even the largest network television advertisers are beginning to use these techniques. The difference in production value between the national network spot and the locally produced spot will begin to disappear.

Figure A-4 Computer graphics make customized, eye-catching commercial messages at very low costs. (Cable Graphic Sciences, Seattle, Washington)

EXAMPLES OF COMPUTER-GENERATED GRAPHICS

A common use of the computer techniques is the preparation of logotypes to be used as part of a television commercial. The screens in Figure A-4 show some logos prepared from digitalized video images and enhanced using the computer techniques, and others produced entirely through the use of tablets and computer instructions.

Appendix B

How the Local Advertiser Gets the Commercial Message on Cable

First, check available cable systems.

> Do they have the facilities for playing commercials?
> Will they help you prepare commercials?

A single telephone call can get answers to those questions.

If the answer is "yes" in both cases, find out:

> How many subscribers does the system have?
> Where are the subscribers located?

If the system reaches the people who are in your sales area, you now have a new way to communicate with them.

Next, check all advertising availabilities on the system. Most will be existing 30-second time segments within programming carried on satellite networks. Do those networks offer programs that will appeal to your prospects?

Some will be on locally originated (LO) shows. Can you sponsor, or partially sponsor, such shows?

You can do messages that will be *read* (alphanumeric) like any printed words . . . or . . . you can use pictures, sound, and music for the full effect of the television screen. Computer graphics are often available to make visuals more exciting.

Which will serve you best?

Once you know what's available and how much it costs, determine exactly how much you will need to spend to reach the audience you want with the advertising you need to convey to them. Almost always, a *campaign* will be required, rather than a single exposure.

If you are sponsoring a program, make sure it is well promoted. Set a budget and a time frame for the advertising.

If it is for a specific event, like a sale, heavy exposure over a short period of time is desirable.

If it is to establish an image, to build a reputation and a continuing clientele, consistent advertising will be needed. An annual budget predicated on a percentage of anticipated sales can be set.

Negotiate price with the cable system. Ask about a long-time contract with free or at-cost commercial production.

Ask if there is co-op money available to you. Manufacturers of merchandise you carry often have such funds available. Most systems are informed about such funds.

Check on *promotional support.* What will the system do to promote the shows and channels on which your commercials appear?

Once you have decided to use cable, you can turn your attention to the commercials themselves.

VIDEOTEX COMMERCIALS

Keep them simple, factual, informative. Make every word count. Tell people exactly what you're selling, what the cost is, when and where to buy it.

Decide how often you want to change the message, and provide copy changes well ahead of time.

Try using other media (program guides, direct mail) to promote viewership of your cable messages. Often the cable system can help you with promotions.

Put some urgency into your advertising. "Sale closes Thursday," "Limited Supply Available." These words invite immediate action by the viewer.

First and foremost, use the selectivity of cable and talk directly and specifically to those particular viewers you've chosen. If it's outdoorsmen, don't have your spokesman in a three-piece suit. Let him wear the clothes and talk the language of cabin and campfire.

If you have certain advertising slogans, symbols, ideas that you have used in other media, carry them over into cable. This enhances all your previous efforts.

Got a jingle or musical symbol you've used on radio? Carry it over into cable, if appropriate. It's a great recognition device. Otherwise, think about establishing music as a background for your commercials. Most systems have music libraries, and they can help you find a theme.

Now concentrate on *what you want to say.*

Set an objective for your advertising. Exactly what do you want it to do?

Make people more aware of your location?

Sell your service capabilities?

Move some particular items of merchandise?
Elicit telephone responses?
Indicate specific price advantages?

Don't try to make every commercial do everything. One simple, clear, action-inducing message is all a 30-second commercial can really accommodate.

If you are buying longer commercials, of course they can go into more detail, but each one should have a specific purpose.

Once you have your objectives clearly stated, you can enlist the assistance of the local system's writing and production staff. Get their ideas of the best way to show and tell your story. A few things to remember:

Good product pictures are important. They should reflect the quality of the merchandise.
Demonstrations of the product in use are persuasive.
Sets, props, lettering, graphics—all reflect on you, the sponsor. If they look cheap and unprofessional, so do you.

You could be the best spokesperson for your business. Take a look at yourself on videotape doing a commercial. Get some objective opinions of your performance. If you come across well, fine. If not, use voice-over pictures or a professional presenter.

After deciding on the general approach and techniques to be used, get a complete script written, with all necessary visual instructions as well as spoken copy. (See examples.)

Select music if the spot requires it.

Have a read-through of the script and make any changes required.

Arrange for a rehearsal. The more you prepare before going into the studio, the less expensive studio time will be required.

If you are planning on using several commercials, make several at one session. That will save you money.

Get started. These advertisers did:

Sir Pizza

In Lansing, Michigan, the pizza business is highly competitive. So Sir Pizza decided to use cable to gain an edge, particularly with the young people who are major consumers. They bought a schedule on Continental Cable, and the system wrote and produced commercials as part of the package. Using a rhythm background and scat lyrics, they brought the following script to life:

TITLE: "Good to the Edge"
Length: 30 seconds

Video		Audio
Snapping fingers against black background.	VOICE:	(with rhythm) Say, folks, don't you know, Sir Pizza has more of what you buy pizza for. . . .
Dissolve to ingredients. Pan them.		Our meats and cheeses are the very best, and our vegetables are super fresh.
Show Royal Feast pizza, complete.		And when you cut yourself a wedge, you'll get pizza that's good to the very edge.
Super phones and phone number.		How do you get the pizza with more? There's fast, free delivery to your door.
Cut to Sir Pizza crest and location.		You can eat it in or take it out, Sir Pizza knows what it's all about. Sir Pizza . . . yaaaaaaaaaa!

Two other 30-second commercials were also written and produced for this advertiser.

Hassan Bros., Inc., AMC

The Cable Advertising Bureau offers a service highlighting examples of commercial creativity. One such commercial was written and produced by Quincy Cablesystems, Quincy, Massachusetts. The purpose was to show a full line of cars, introduce the sales staff, and establish the location. Cost to the client was $400. Production time was a week and a half.

TITLE: "What's the Best Way to Get There?"
Length: 60 seconds

Video	Audio
Open with front shot of J. Hassan driving into lot. Hassan Bros. sign is visible in background. Camera follows as he pulls into parking spot. SUPERIMPOSE price over car.	(Music upbeat and friendly) ANNOUNCER: Meet John Hassan of Hassan Bros. AMC in Quincy. John loves showing off the AMC Alliance. It was selected car of the year by Motor Trend magazine. . . .
Cut to shot of Encore entering picture, pulling up to Alliance, and parking next to it. Shot also includes D. Hassan lifting hatchback and removing brown lunch bag. SUPERIMPOSE price over car.	. . . John's brother David loves to show off the Encore. He likes it because of the hatchback. . . .
Cut to front shot of Eagle heading to camera. Camera follows car as it parks next to Encore. Driver opens car's door and stands there. SUPER price over car.	. . . And this is Willard Dunlap. Willard gets excited showing off the Eagle. He lets everybody know it is the only four-wheel-drive passenger car built in America . . .
Cut to front shot of CJ7, which comes bouncing into lot and parks next to the Eagle. Driver gets out. SUPER price over car.	. . . Donald Hassan is always having fun showing off the CJ7 Jeep. . . .
Cut to front shot of XJ Wagoneer entering lot. Camera trucks right to include side shot of car. SUPER price over car.	. . . And this is Donald's father, Turk Hassan. Turk prefers a luxury four-wheel-drive wagon. He's proud showing off the XJ Wagoneer. . . .
Cut to long shot as camera pans line of 5 parked cars. Everybody seen walking to showroom entrance.	. . . With such a great line of automobiles, the boys have a hard time agreeing on which car they like best. One thing they do all agree on is that Hassan Bros.

Video	Audio
	offers the lowest prices on the South Shore. . . .
Cut to Hassan Bros. as they are about to enter showroom. They all turn to camera and attempt to give directions in response to announcer's question. They are all pointing in opposite directions.	. . . By the way, fellas, what's the best way to get there? (Hear arguing as they all point to different directions.)
Camera zooms out, showing Hassan Bros. sign. The brothers are arguing with one another.	Well, you'll find Hassan Bros. at 290 Washington Street, just outside of Quincy Square.

Community National Bank

The objective was to dramatize the area served by this financial institution and to highlight some of the services offered. It was felt that a two-minute infomercial could best do the job. Another assignment was the identification of the specific communities where the bank had offices. This was accomplished with a carefully designed animated logo. Locations were shot on film, which was transferred to videotape for editing and finishing.

TITLE: "Meet Community"
Length: 120 seconds

Video	Audio
Copter shot of fall foliage, lakes, and homes in Bloomfield Township. Dissolve to Silverdome from air.	(Sound of copter engine) MARK: Take a look at the communities served by Community National Bank (sound fades out) and see if you're familiar with some of the landmarks. . . .
Dissolve to shot of Oakland University's Meadowbrook Mansion. Christine Schaeffer and several students walk by. One turns to go in house. She's car-	Like Meadowbrook at Oakland University. Our University office manager knows that campus well.

Video	Audio
rying a bouquet of fall leaves. Waves before entering, others exit.	
Cut to proving-ground track.	Not far from another Community National Bank, there's the proving ground.
Cut to Milford Branch executive driving past proving ground.	It's right on the way to our Milford headquarters.
Beautiful apple orchard and old house in Romeo. . . . Harry Bennett and the owner on lawn, talking. A hunting dog comes into picture, runs to the men. Bennett pats the dog.	Have you seen the lovely apple orchards and old homes of Romeo? They're a familiar sight to Community National's local manager . . . and he's a familiar figure around town.
Clarkston Mills . . . Betty Retter chats with a shop owner, or perhaps Marc Alan, the developer. They are watching landscapers at work.	Just a few blocks from Community National in Clarkston is the exciting new Clarkston Mills Shopping Center. Here our Clarkston manager and a new neighbor watch work in progress.
Lake Orion . . . at a nice waterfront area . . . John Gardiner and friend.	A nice place to discuss business . . . along the shore line of Lake Orion. Community's Lake Orion manager explains the new simple interest loan plan to a customer and friend.
Avon Township Library . . . Japanese tea garden . . . Bob Shuller and an elderly couple. They are carrying travel literature, showing it to Shuller.	While in front of Avon Township's library, the garden is a pleasant place to meet. The man from Rochester's Community National explains the advantages of MasterCharge and Visa arrangements to some long-time customers about to take a long-awaited trip.
Farm picture . . . near Washington Township. Ron Davis talking to a father and son, both farmers; they're standing by	In Washington Township, our office manager describes the automatic funds transfer plan that protects your checks by making

Video	Audio
some new, big farm machinery.	sure they're covered.
Cut to long shot of Pontiac Mall. Cut to several Community executives in conference room, considering installation of ATM.	At Pontiac's busy mall, and in other locations, there will soon be automated teller machines. Here, Community National executives plan the new Bank Anytime System.
Fast cuts . . . Kirk of the Hills . . . Walled Lake Stonecrest Building . . . Waterford's old mill. Branch bank shots are intermixed with Union Lake Volunteer Fire Dept., Cass Lake Marina, Wixom-Ford Plant, Silverdome, White Lake Alpine Ski Lodge. When possible, bank personnel are in pix. Maybe some bank interior transaction shots can be included; teller's window, desk discussion, someone entering safe deposit vault.	You see, wherever you live in Oakland or Macomb County, Community National Bank is out to prove that their interest . . . (pause) is in you . . . in making your money work harder for you . . . in providing better services at reasonable costs to you.
Cut to downtown Pontiac headquarters. Pull back to show name above door, then pan down to Pete Eierman and two other men on steps, talking.	That goes for everybody, from the chairman . . . to the hundreds of people at the 28 offices of Community National.
Cut to copter shot . . . logo over. Hold logo as we write on Community names: Avon Twp., Bloomfield Hills, Commerce Twp., Independence Twp., Keego Harbor, Lake Orion, Milford, Pontiac, Pontiac Twp., Romeo, Royal Oak Twp., Walled Lake, Washington Twp., Waterford Twp., White Lake Twp., Wixom.	Come in tomorrow and talk to your Community banker. Community National Bank . . . *where the interest is in you.* Member FDIC.

Courtesy Community National Bank.

Smaller banks around the country might consider this sort of locally related advertising as a means of competing with the growing number of huge banking systems that are moving into small towns and rural areas.

REAL ESTATE

Another business group that should consider cable's possibilities is the real estate sales operations looking for a competitive edge.

What are real estate firms now doing with cable in their communities? Consider the advertising profiles of some 40 cable systems throughout the country. One portion of that profile is a statement by the cable system as to its "best ad accounts." Listed regularly are auto dealers, insurance companies, financial institutions, hotels, and department stores. Conspicuously absent are real estate firms. But there are some notable exceptions:

> In Seattle, the cable system carries a one-hour show each night from local realtors, called an "Electronic Listing Service," in which homes are featured in considerable detail, using footage created in much the same way I have suggested you attempt.
>
> In Phoenix, the cable system recently introduced the Home Channel as 24 hours of videotaped informational advertising about homes and services associated with home ownership, such as plumbing, heating, and remodeling.
>
> In Lansing, Michigan, the cable system has introduced "The Hunted House," one hour each day of local homes on the market, supplemented also by informational packages concerned with such topics as tree removal, solar energy options, home tax credits, and so on.

Because the typical cable system is just beginning to get into the advertising business, it is likely that many real estate firms have not been contacted as potential advertisers. Often it is necessary to initiate some discussion with cable system managers. Expressed interest in using cable as a further advertising vehicle may put the aggressive real estate firm into a particularly strong position to negotiate price and placement of ads. You may be surprised to find the system eager to try some experimental advertising attempts with your firm at extremely low costs initially.

Why think about advertising by cable? First is the opportunity to get unlocked from thinking in terms of 30-second or 60-second commercials. Cable has the flexibility to use virtually any time unit that suits the product. Clearly, with real estate advertising, all that can be done in a typical spot commercial on traditional television or radio is to identify the firm and perhaps feature a single home. With cable, you can design and use commercials of virtually any length, commercials that are as informative about available property as you wish them to be. Featuring a half-dozen homes in considerable detail, visual as well as audio, in a half-hour sequence could be a regular format. The point is that the format is flexible in terms of time length, in terms of kinds of information, and in terms of mode of presentation.

The price is still very, very right. Consider a cable system that now services more than 60,000 subscribers. Leasing 30 minutes on one of the cable system's local origination channels would cost an advertiser $100. On smaller systems, in the 20,000–25,000 subscriber range, it would cost $50.

An advertiser can choose the time or times for the advertising to appear. Channels available for this purpose are not yet saturated. When would potential buyers be most interested and most available to view such information advertising? Saturday afternoon before Sunday open houses? Sunday noon? The advertiser can usually pick the time for the placement of most of cable advertising.

Messages will be available in a specifically defined geographic location. This is the concept of narrowcasting, casting a smaller net but one that is more definable, in contrast to the broadcasting net of commercial radio and television that wanders into and across geographic areas that may be of little interest to the advertising company.

The cable system is likely to have market research information about its subscribers—who they are, what their purchase patterns have been, what their viewing interests are, when they view, family composition, and other demographic information. All of this can assist you in deciding how and when to approach them with real estate information. If the cable system does not have much market information, insist that they obtain it, if they wish you to become an advertiser. Nationally, there are good data available to indicate that the prime subscribers to cable are younger, married adults, with children and an upscale income and life-style. These characteristics also fit those cable subscribers who are most prone to take on pay channels as part of their cable investment.

A further advantage in thinking about cable advertising is the flexibility in formats available. A typical cable system, in addition to carrying network television and at least one public television station, also carries a variety of satellite networks—for example, a sports network, an all-news network, a religious programming network, a children's programming network, and so on. It provides several channels of automated text—for example, business news, stock market news, sports news, a weather channel. It provides local origination channels. It provides local access channels that often carry high school sports, city council meetings, local 10-kilometer races, and other community activities. It may already provide shopping channels. And it has one or more premium or pay channels. With the exception of the commercial and public television networks and the pay channels, the remainder of the offerings are available for advertising messages.

At present, most local cable systems can plug in two minutes of advertising on ESPN, CNN, USA, and other satellite networks, including MTV. Spot advertising on such networks costs about $1 per 1,000 subscribers, or $50 for a 30-second spot in a system with 50,000 subscribers. That price will be substantially less than a commercial television station serving the same community will charge, and quite comparable to broadcast radio advertising on the top-rated stations in town.

SUMMARY

Thirty seconds, 60 seconds, 120 seconds to tell a story that can influence your prospective customers. Sight, sound, action, music, or simply printed words passing on the screen. Choose your length and technique. Find exactly the system and show that can best reach your potential market.

Use cable the way it works best, as a *new* medium with attributes all its own.

Appendix C

A Simple Audience Research Project

The most cost-efficient way to do a local audience research project is through a telephone survey. A mail survey is impractical because of probable low completion rates and the lack of statistical and analytical capability necessary to handle diary data. A personal-interview survey simply requires too much personnel time. A telephone survey fits the resources available to most cable operators or advertisers—that is, a large number of telephone lines. Interviewing can also be done during off-business or evening hours, when most of the telephones should be idle. The operator can use internal personnel or employ temporary labor.

The best method to measure cable audiences is the method known as telephone recall, a method used by commercial firms to measure network radio audience and television commercial audiences. The respondent, or person being interviewed, is read dayparts, usually for about four-hour time segments, for either yesterday or today, and then asked from memory to reconstruct his or her viewing. Generally, people cannot remember their viewing patterns for more than one day accurately enough to report them reliably. This means that the survey needs to be conducted on different days if more than one day is to be included in the study. Probably the most convenient scheme for most cable operators is conducting interviews in the evening, using yesterday's programs.

DRAWING A SAMPLE

The first step in conducting this research project is to obtain the sample. First, the size of the sample needs to be determined, and then the sample members need to be identified. Avoiding a complex discussion of statistical sampling theory, a good rule of thumb would be that no group to be used for the calculation of a rating in the numerator should be less than 30 individuals. This means that if a particular channel is expected to be viewed on the average by 5 percent of the subscribers, a sample of 600 would be necessary, because 30 is 5 percent of 600. If a channel would be viewed by 1 percent, then a sample of 3,000 would be necessary. A sample of 1,500 could be obtained from interviewing 300 each night of the week. Three hundred interviews in an evening would require between ten and twelve interviewers for approximately four hours. If this effort is too great to be practical for the researcher, then the interviewing could be conducted across several weeks, until the desired number of interviews is completed. If, for example, only six interviewers in an evening are available, then it would require two weeks to complete the study.

The best way to select the sample is to begin with a complete list of cable subscribers. Some care is necessary to make sure that the list of subscribers does not contain any geographical bias—that is, arrangement of the list according to neighborhoods. An alphabetical list with telephone numbers would obviously be the best. If telephone numbers are not part of the list, then they will have to be obtained from the local telephone directory. The first step is to determine the number of subscribers needed in the sample. This is done from the estimate of the sample size, as discussed before, adjusted for completion rate and other factors. If telephone numbers are not readily available and the local directory must be relied on, it would be necessary to have an estimate of the proportion of usable telephone numbers. In some larger metropolitan areas, as high as 40 percent of the numbers in a telephone directory are not usable. In smaller and more stable communities, the proportion of usable numbers may be very high and may not be a factor at all. Not every subscriber will agree to be interviewed, nor will all be available to be interviewed during the study. A good estimate of the proportion of subscribers that would fit into this category would be 30 percent. This means that to ensure 1,500 completed interviews, an initial sample of approximately 2,150 would need to be drawn. The total initial sample estimate can be calculated by dividing the total number of interviews desired by the completion proportion—in this case, 70 percent, or 0.7.

Two numbers are now important: the size of the initial sample and the total number of subscribers. A skip interval can be determined by dividing the size of the initial sample into the total subscriber base. If there are 50,000 subscribers in the system and an initial sample of 2,150 is needed, then the skip interval would be approximately 23. The skip interval means that every 23rd name would be selected for the sample from the complete list of all subscribers. One other number is also necessary: the starting point before the first interval. Using some random procedure, such as rolling dice or drawing numbers out of a hat, a starting point must be chosen. If the number selected is 11, then the eleventh name on the list would be the first subscriber chosen for the sample. If the skip interval is 23, then the next name chosen would be the 34th name, then the next would be the 57th, and so on.

THE QUESTIONNAIRE

The second step in the process is the development of a questionnaire. The questionnaire should be simple and straightforward, so that it is easy for relatively untrained interviewers to administer. Initially, the questionnaire should begin with questions that identify the appropriate respondent within the household. The first question should be simple and asked immediately after the introductory greeting. The object is to get the respondent involved before he or she can say "no." Questions like, "Are you a cable subscriber?" should be asked, as well as, "Did you view television yesterday?"

Additional screening questions and instructions may be necessary to ensure representation of men in the sample. Normally, an adult female member of the household will answer the telephone in three out of four calls. If adult males are never asked for, they would of course be underrepresented in the sample. A solution to this problem is to always ask for an adult male member of the household. If the adult male is not available, then the interview would be conducted with whoever answered the telephone. This results in an approximately equal distribution between males and females and, if desired, an appropriate representation of teen-aged children. If children under 12 are desired, a different type of screening sequence would be written.

The questionnaire would then proceed in a most straightforward manner, with yesterday's roster of dayparts available to the cable subscriber. For example, the interviewer would begin with the morning hours and follow with the afternoon hours, the early evening hours, and the rest of the day. Then the respondent is asked to reconstruct his or her viewing habits, reporting the channel, the starting time, and the ending time. After the viewing habits have been reconstructed, a few simple demographic questions should be asked. Most important are age and gender, but also useful might be income and occupation. A sample questionnaire is shown at the end of this appendix.

DOING THE SURVEY

The third step, of course, is actually administering the questionnaire and collecting the data. It is important that the proper attitude toward the survey is instilled in the interviewers. This is usually done in a training session, where the importance of the study is noted and practice interviews are conducted among the interviewers. Important to stress in the training session are accuracy in the recording of responses and avoidance of leading respondents in their answers. Obviously, interviewers should not ask questions like, "You did watch USA network, didn't you?"

Each questionnaire—and there should be one questionnaire for each interview—should be checked and numbered upon completion by someone other than the interviewer, to help eliminate error and ensure high-quality results. Checks should be made so that the interviewer's writing is legible and complete. Numbering the questionnaires makes each one easily identifiable in the later analysis stage.

PREPARING THE REPORT

The next stage involves the tabulation and analysis of the data collected. Since most cable operators would not have the computer capability to machine-tabulate the results, the best way would be to tabulate them by hand. This can be efficiently accomplished by using large spread sheets that have the channels across the top and the time of day down the side of the page. A spreadsheet with rows and columns will facilitate the analysis of any demographic subgroup. "Men 18–49," "Women 18–49," and "All adults" are useful categories. The times that label the rows on the sheet would be divided into appropriate time segments. Using some rule, such as six minutes within the 15-minute or quarter-hour segment, an individual respondent would be given credit for viewing the channel according to the data recorded on the individual questionnaire. This would be done by making a mark or a 1 in each appropriate box and then proceeding to the next questionnaire. Also useful would be a "total" box that would record the total number of respondents in that demographic category to make possible the computation of rating percentages.

An electronic spreadsheet program, such as those found on many personal computers, could be used for this task as well. After all the data have been entered on the appropriate worksheet, totals can be derived from each quarter-hour or longer time segment throughout the day. Rating percentages can be computed directly from these totals. Appropriate aggregation can be made across the various demographic groups to arrive at total audience estimates. Also, averages or summaries for individual programs can be determined. Another important number that should be kept on the spread sheets is the number of subjects who tune into any programming service anytime during the day. These numbers would probably be most easily kept at the bottom of the spread sheet for each appropriate program column. These totals would give the cumulative audience for that programming service for the day. This could be compared to the average quarter-hour audience calculated for the same programming service.

Finally, it is important that the results of this survey be presented in a form that is easy for the cable sales staff and advertisers to use. A booklet or research report that is also simple and straightforward is vitally important. The report should include the daypart estimates as well as the cumulative estimates for all the services that the cable operator has advertising availabilities to sell. If the cable operator is selling packages across several services, it might also be a good idea to keep a cumulative audience count on those services that are packaged, to provide a cumulative audience estimate for all the services combined. For example, if CNN and the USA network are sold together, then an advertiser would be interested in the total audience reached by the two services together. This could be shown on the research report as a special package page.

A research report should be attractively laid out and prepared. Of prime importance would be a few pages describing the methodology employed in conducting the research, including a description of the sample, a report of the interview times, and a summary of the sample characteristics.

A report like this should provide very useful information for the cable advertising sales staff. The numbers provided are equivalent to those of the other broadcast competitors, including both local television and local radio. They should also counter the often-heard argument from advertisers that cable does not have numbers.

AUDIENCE STUDY INTERVIEWER # ______________

CASE # ______________

TODAY IS: __MON. __TUE. __WED. __THU. __FRI. __SAT. __SUN.

DATE: ______________

Hello, I'm __________ calling from XYZ Cable Company and we're doing a survey about television. (If a woman answers) I'm having trouble getting enough men for my survey and I'd like to speak to an adult male if there is one there. (If no man present) I'd like to ask you the questions then.

1. First, do you live in the City of Anywhere?
   _____ YES (GO TO Q. 2)
   _____ NO (TERMINATE)
   _____ DK/REFUSED (TERMINATE)
2. Since our survey is about television, I need to know how many television sets you have in your household.
   (RECORD NUMBER. IF NONE, TERMINATE) ______________
3. Do you subscribe to cable television?
   _____ YES (GO TO Q. 4)
   _____ NO (TERMINATE)
   _____ DK/REFUSED (TERMINATE)
4. What tier of service do you have?
   _____ Basic Service
   _____ Enhanced Service
   _____ DK/REFUSED
5. What premium services do you have? (CHECK ALL THAT APPLY)
   _____ HBO
   _____ Showtime
   _____ Movie Channel
   _____ NONE
6. I would like to ask you about the time you spent watching television yesterday, that is (DAY OF WEEK), at home. Think about the time period from (DAYPART) yesterday. Did you watch at all during this time period?
   DAYPART

Midnight up to 6:00 A.M.	_____ YES	_____ NO
6:00 A.M. up to 3:00 in the afternoon	_____ YES	_____ NO
3:00 P.M. up to 6:00 in the evening	_____ YES	_____ NO
6:00 P.M. up to midnight	_____ YES	_____ NO

7. (ASK FOR EACH "YES" DAYPART) Now, from the time you were watching, what channels or stations did you watch from (DAYPART)? (RECORD STATION OR CALL LETTERS ON THE "VIEWING RECORD," PROBE IF NECESSARY) Well, can you give me the number of the channel on the dial? (IF STILL NO RESPONSE, ASK FOR SLOGANS OR OTHER IDENTIFICATION)

   Did you watch any other channels or stations from (DAYPART)? (CONTINUE UNTIL DAYPART COMPLETE)

8a. What time did you start watching (CHANNEL) (first/again)?
8b. Did you begin exactly at (TIME MENTIONED) or did you begin a few minutes before or after that time?
8c. (IF BEFORE/AFTER, ASK) About how many minutes (before/after)?
8d. And what time did you (first) stop watching (CHANNEL)?
8e. Did you stop exactly at (TIME MENTIONED) or did you stop a few minutes before or after that time?
8f. (IF BEFORE/AFTER, ASK) About how many minutes (before/after) did you stop watching?

REPEAT Q. 8a–8f UNTIL DAYPART IS COMPLETE.
WHEN DAYPART COMPLETE, GO TO Q. 7 FOR NEXT DAYPART.
Now, just a few questions about yourself.

9. Do you have any children under age 12 in your household?
   _____ YES
   _____ NO
   _____ DK/REFUSED
10. Would you please tell me your age?
   (RECORD TO NEAREST YEAR, 00 FOR REFUSALS) ___________

Thank you very much.

11. RECORD SEX: _____ FEMALE
   _____ MALE

VIEWING RECORD

TIME BEGAN		TIME STOPPED		CHANNEL
______	am noon pm mdnt	______	am noon pm mdnt	____________
______	am noon pm mdnt	______	am noon pm mdnt	____________
______	am noon pm mdnt	______	am noon pm mdnt	____________
______	am noon pm mdnt	______	am noon pm mdnt	____________
______	am noon pm mdnt	______	am noon pm mdnt	____________
______	am noon pm mdnt	______	am noon pm mdnt	____________
______	am noon pm mdnt	______	am noon pm mdnt	____________
______	am noon pm mdnt	______	am noon pm mdnt	____________
______	am noon pm mdnt	______	am noon pm mdnt	____________
______	am noon pm mdnt	______	am noon pm mdnt	____________
______	am noon pm mdnt	______	am noon pm mdnt	____________
______	am noon pm mdnt	______	am noon pm mdnt	____________
______	am noon pm mdnt	______	am noon pm mdnt	____________
______	am noon pm mdnt	______	am noon pm mdnt	____________
______	am noon pm mdnt	______	am noon pm mdnt	____________
______	am noon pm mdnt	______	am noon pm mdnt	____________
______	am noon pm mdnt	______	am noon pm mdnt	____________
______	am noon pm mdnt	______	am noon pm mdnt	____________

Glossary

Certain words and phrases in this book are often used in discussion of cable and/or television advertising.

Access channel Many cable systems reserve such channels for use by the public and public institutions. May make underwriting possible for advertisers. Cable operator has no editorial control.

Advetorial A word coined to describe commercial messages that present a point of view, usually institutional.

Affiliate A TV station or cable system that has a contract with one of the three major networks or one of several cable networks to carry its programs.

Audio origination Audio signals that originate at the cable system headend.

Automated channel One that automatically repeats text messages.

Availabilities The inventory of commercial time available for sale.

Barter A deal where the advertiser exchanges his goods or services for commercial time on a program.

Basic cable TV A minimum number of channels available for one basic price to the subscriber, the first tier of service.

Bonus Extra spots that are negotiated into a contract for the advertiser or given if availabilities are not sold.

Bulk plans A series of advertising purchase plans or packages based on a volume discount structure.

CAB The CableTelevision Advertising Bureau, Inc., 767 Third Avenue, New York, New York 10017, (212) 751-7770.

Cablecasting Transmission over a cable system that reaches only that system's subscribers.

Cable radio FM service or audio origination (see separate definitions) or an audio background to accompany a silent automated text channel.

Cable system A franchised operation reaching a specific area via cable.

CATV The original designation for cable, Community Antenna Television.

Computer graphics Inexpensive, high-tech way of achieving artistic effects in shows and commercials.

Continuity Consistency of advertising placement through time.

Continuity discount Discount based on buying a certain number of spots for a minimum number of weeks.

Cost per thousand (CPM) Cost to get one advertising exposure to 1,000 persons.

Crawl Moving text across a section of a TV screen for continuous reading.

Data channels Those that carry constant information in text and graphics; i.e., weather, news, financial information.

Daypart Division of the broadcast day by time blocks.

Demographics Classification of viewers into categories such as sex, age, income.

Direct TV selling (direct response) Consumers order directly in response to a television advertisement by writing or calling an 800 phone number. May also refer to shopping channels and interactive TV purchases where consumers are automatically charged for what they use or transmit.

Dual rate structure One rate for national advertisers and a second, lower rate, for local advertisers.

Effective reach Number of exposures necessary to achieve the objective for a campaign.

Fixed placement Buying commercial time in specifically designated programming or time slot.

Flighting Advertising for a short period of time and then not at all.

FM service Local and distant FM signals provided by a cable system as part of the basic package or as an additional service for a monthly fee.

Franchise A form of contract between the cable TV operator and the municipal authority.

Frequency The average number of times a prospect who is reached is exposed to a message.

Frequency discount A discount applied on the basis of the number of spots purchased within a specified time interval (e.g., one week).

Grid Various classes of time or programs that are representative of different audience levels.

Gross rating points Total audience, including people or households exposed more than once.

Hard interconnect Cable systems physically interconnected for purposes of selling advertising.

Hardware Equipment used in a cable system.

Headend Control center and signal source for the cable system.

Horizontal rotation Purchase of advertising at the same time across different days.

Infomercial A long commercial with a great deal of information about the product or service and its use.

Insertion equipment Computer-controlled videotape machines that automatically play advertising messages at the proper time on the proper channel. May be cued by signals from national cable network transmission codes. A printer with the insertion equipment documents the day, time, and channel of play for the cable system and advertisers.

Institutional advertising Messages designed to enhance a public image, provide information, or promote a particular idea. No attempt to directly sell a product or service.

Integrated commercial One that is an integral part of the show, often using the same talent as that used in the entertainment.

Interactive TV Two-way TV service by which the subscribing household can transmit "upstream" to the cable headend. May be used for ordering information, as in videotext, or ordering merchandise, as in an interactive shopping channel.

Interconnect An arrangement linking two or more cable systems so that time can be bought through a single source. Creates a regional media buy.

Inventory All the available advertising spots the cable operator has to sell.

Local origination (LO) A program that originates locally, over which the local cable system and sometimes advertisers have editorial control.

Local spot A commercial-directed availability sold to a local advertiser interested in reaching viewers in a limited geographic area.

Merchandising Extra help the cable operator may give the advertiser, such as point-of-purchase display materials, promotional announcements of sponsored programs, guide advertising, etc.

Multiple-system operator (MSO) A company that owns more than one cable system.

Narrowcasting Targeted programming on cable aimed at a highly select audience.

National spot A commercial purchased from among local availabilities by a national or regional advertiser.

Network advertising Advertising in network programming transmitted to the entire network.

Nielsen rating A rating system used by advertisers to evaluate viewership of various shows. Includes cable networks reaching more than 15 million subscribers.

Package (see *Plan*)

Placement Where spots are located in the cablecast day or in the program schedule.

Plan The purchase of a multiple of advertising spots permitting discounts.

Preemptible If advertising time can be sold at a higher rate, the higher-paying advertiser gets the time. The advertiser takes a chance on "preemptible" time to get a better rate.

Prime time When most viewers watch TV, usually between 8 and 11 P.M. Could be defined differently in different communities.

Promercial A commercial promoting a particular event. Often longer than 30 seconds.

Psychographics Classification of viewers by life-style.

Pulsing Periodically increasing advertising while always maintaining a minimum level.

Rate card The specific charges for advertising time on each cable system.

Rating Estimated percentage of all TV households or people in a specified area tuned to a particular program at one time. Expressed as a percentage of the total potential audience.

Regional spot A commercial intended to reach a fairly broad geographic area, usually involving several cable systems.

Remotes Material, either for programs or commercials, that is shot at a location away from the studio.

Roadblocking Using all available commercial time for a single advertiser at a particular break in programming on all available channels.

ROS (run of schedule) A designation that means commercials can be run whenever the system has a time slot available. Sometimes the contract calls for a specific number of each time slot on each channel involved in the purchase.

Satellite channel A TV channel beamed to an earth station (dish) from a satellite and modulated on a cable system.

Satellite network An organization that controls the programming on a particular channel and beams it to individual cable systems via satellite (such as MTV, the Music Network; CNN, the Cable News Network).

Segmenting Differentiating consumers and audiences so that marketing strategies and advertising messages may be specifically targeted.

Share The estimated sets in use tuned to a specific program, expressed as a percentage of all households or people viewing at a particular time.

Soft interconnect Cable systems interconnected for purposes of selling advertising, but not physically (tapes are delivered to each system).

Software The programs appearing on cable systems. In cable advertising, also used to describe computer programs that control computer graphics.

Sponsorship The purchase of an entire program segment and control of all commercial time within it.

Spot advertising Advertising purchased separately from a single cable system or several selected cable systems.

Superstation A broadcast station that is retransmitted by satellite to many cable systems. All advertising is carried on the original broadcast. There are no local insertions.

Tracking Following up advertising campaigns to attempt to determine results.

Underwriting Support for a program without a direct sales message. A statement of funding support before and/or after the program identifies the benefactor and earns a certain amount of public favor. Used on public broadcasting and a possibility for access channels and local origination.

Upscale Households or persons of above-average income.
Vertical rotation Purchase of advertising at different times within the same day.
Videotex A graphic display system of printed words and graphics. Accessed at the command of the cable consumer for information and advertising.
Volume discount (see *Frequency discount*)
Wearout Placing too many advertising exposures against a target audience.
Zapping Switching away from a commercial with a remote tuner or deleting commercials from recorded programs.

SOME BASICS OF CABLE TECHNOLOGY

A cable system has three major components: the headend, the distribution plant, and the home drop.

The *headend* is the control center of the system. It receives broadcast TV and radio signals off the air from antennas on a high tower. Some of the most distant stations may come in to the tower by microwave. At the base of the antenna are several large earth stations pointing at the satellites servicing the cable industry. Inside the headend building are studios for the television production of locally originated signals, character generators for text, commercial insertion equipment, and receivers, processors, and modulators that put the signals from the various sources on the cable.

The *distribution plant* is called a tree system. The *trunk* of the system goes down all the main arteries of the community. Branching off are *feeder lines* that serve residential streets. Amplifiers are placed in trunk and feeder lines to keep boosting up the signal as it attenuates.

The *home drop* taps off the feeder lines to go to each household. A converter in the home is used for selection of channels while the set remains tuned to a single prescribed channel. The sophistication of the converters varies. Some may be addressed from a computer at the headend with the various levels of service offered (basic and the several pay channels). For the nonaddressible converters, whenever a program service is added or discontinued, a service call must be made and a different converter installed.